On the Properties of Novel Superconductors

Edited by Heshmatollah Yavari

Published in London, United Kingdom

IntechOpen

Supporting open minds since 2005

On the Properties of Novel Superconductors
http://dx.doi.org/10.5772/intechopen.81007
Edited by Heshmatollah Yavari

Contributors
Hong Yi Chen, Gulshan Prakash Malik, Vijaya Shankar Varma, Ekaterina Kurbatova, Pavel Kurbatov, Mikhail Sysoev, Rafael Zamorano Ulloa, Heshmatollah Yavari

First published in London, United Kingdom, 2020 by IntechOpen
IntechOpen is the global imprint of INTECHOPEN LIMITED, registered in England and Wales, registration number: 11086078, 7th floor, 10 Lower Thames Street, London, EC3R 6AF, United Kingdom
Printed in Croatia

British Library Cataloguing-in-Publication Data
A catalogue record for this book is available from the British Library

Additional hard and PDF copies can be obtained from orders@intechopen.com

On the Properties of Novel Superconductors
Edited by Heshmatollah Yavari
p. cm.
Print ISBN 978-1-78984-065-0
Online ISBN 978-1-78984-066-7
eBook (PDF) ISBN 978-1-83968-441-8

For EU product safety concerns:
IN TECH d.o.o., Prolaz Marije Krucifikse Kozulić 3, 51000 Rijeka, Croatia, info@intechopen.com or visit our website at intechopen.com.

Meet the editor

Dr. Yavari is a Professor of physics at the University of Isfahan. His research interests are superconductivity, superfluidity, supersolidity, Bose-Einstein condensation, ultracold atomic gases, optical lattices, granular electronic systems, neutron stars, topological insulators and superconductors, Majorana fermions, photonic crystals, low dimensional systems, and mesoscopic systems He has published more than seventy papers in journals and conference proceedings.

Contents

Preface

To unravel the nature of the emergent phenomena in strongly correlated many-body systems is one of the central challenges in theoretical and experimental research. Unexpected and exotic properties of these systems open active areas of study in the field of condensed matter physics. One of the remarkable phenomenon of these correlated systems was the discovery of superconductivity by Onnes in 1911. Since then, a lot of experimental and theoretical research has done to describe this phenomenon. The discovery of modern superconductors, such as high-Tc, organic, two band, and some others, presented another criticism for the theoretical physicist. The possible superconductivity mechanism of these systems remains one of the most challenging problems in physics.

Most researchers believe that their superconductivity mechanism could not be explained within the BCS theory. At present, there is no mechanism to describe all the properties of these superconductors. Many different models such as the exciton model, plasmon model, bipolaron model, magnon model, spin fluctuation model etc., have been proposed. All of these models are evaluated by presenting new theoretical efforts and experimental results.

In this regard, this book includes valuable theoretical and experimental discussions about the properties of superconductors. Here you will find valuable research describing the properties of high-T_c and iron-based superconductors.

The main topics of the book are organized in five chapters, as described below.

Chapter 1 contains a brief discussion about the progress in superconductivity.

Chapter 2 discusses the experimental techniques to measure microwave absorption.

Chapter 3 deals with the electromagnetic properties of bulk high-temperature superconductors (HTS), which can be used in magnetic systems for various applications.

Chapter 4 focuses on the generalized BCS equations to discuss high T_c superconductors.

Chapter 5 covers the study of iron-pnictides using self-consistent Bogoliubov-de Gennes equations.

I hope this book is useful for theoretical and experimental researchers.

Heshmatollah Yavari
University of Isfahan,
Iran

Chapter 1

Introductory Chapter: Superconductivity in Progress

Heshmatollah Yavari

1. Introduction

Describing the behavior of strongly correlated systems is a challenge for every area of physics. The many-body ground state that emerges in these systems can have exotic and unexpected properties. As a remarkable of these correlated systems was the discovery of superconductivity by Onnes in 1911 [1], which opens active areas of study in the field of condensed matter physics.

This state is a quantum phenomenon that is displayed by certain materials under particular magnetic and temperature regimes. The Meissner effect discovered in 1933 shows that superconductivity is more than just the disappearance of resistance and it is a true thermodynamic state of matter because it leads to the superconductors as perfect diamagnetism materials [2]. This effect distinguishes a superconductor from a perfect conductor.

To describe these phenomena, a lot of theoretical works were proposed. The first phenomenological theory which is known as London theory was proposed by the London brothers [3]. This theory described well the magnetic properties and magnetic penetration depth of Type I superconductors in weak magnetic fields. By extending the London theory, Ginzburg and Landau proposed a phenomenological theory, which enables the study of Type II superconductors in a strong magnetic field [4]. By assuming the second-order phase using physical intuition and using variational principle of quantum mechanics, the Ginzburg-Landau (GL) theory allowed the calculation of macroscopic quantities of the material in the superconducting state. The most beautiful and successful theory in condensed matter physics, i.e. the microscopic theory of superconductivity which is known as BCS theory was formulated by Bardeen et al. [5].

The starting point of this theory was an effective Hamiltonian of fermionic quasiparticle excitations that interact via a weak phonon-mediated potential close to the Fermi surface. This attractive interaction leads to bound electron pair states known as Cooper pairs. While the single electrons are fermions, not being able to occupy the same state due to Pauli exclusion principle, Cooper pairs are no longer obliged to obey the Fermi-Dirac statistics, and like bosons, they can enter the same state. Superconductors that can be described by the BCS theory are known as conventional superconductors.

The discovery of high-temperature superconductors (HTS) in 1986 [6] presented another challenge for theoretical physicist. Their superconductivity mechanism remains one of the most challenging problems in physics. It is believed that the BCS theory alone cannot explain properties of these classes of superconductors and new theoretical concepts will be required. This unexpected discovery opened a new area in the history of superconductivity, and experimental researchers started trying to find new compounds in this class of superconductors. In the following

 IntechOpen

years, many different HTS compounds with high transition temperatures were found [7–9]. Unfortunately, most HTS compounds are ceramic in nature and thus their industrial application (e.g., cables and conductors) is difficult. The symmetry of the Cooper pairs in exotic superconductors like cuprates, heavy-fermion [10, 11], organic superconductors [12], and Sr_2RuO_4 [13] is different from conventional superconductors. Since the pairing symmetry of these superconductors is no longer s-wave, they are known as unconventional superconductors and the study of them has become a central issue in condensed matter physics. These superconductors are often characterized by the anisotropic character in the superconducting gap function with nodes along a certain direction in the momentum space [14, 15]. Since the pairing interaction has an important role in the superconducting gap structure, its determination is very important to explain the basic pairing mechanism. The experimental techniques for the determination of the nodal structure such as phase-sensitive experiments [16] and angle-resolved photoemission spectroscopy (ARPES) [17] are suitable for HTC cuprates but are difficult to apply to other unconventional superconductors with low transition temperature T_c. Therefore, the superconducting gap structure of most unconventional superconductors is still unclear.

The classification of the superconducting order parameter is determined by its behavior under symmetry transformations. From the viewpoint of symmetry, the full symmetry group of the crystal includes the gauge $U(1)$, crystal point G, spin rotation $SU(2)$, and time reversal symmetry T groups, and has the following form:

$$\mathfrak{R} = U(1) \otimes G \otimes SU(2) \otimes T \tag{1}$$

Below the transition temperature, the $U(1)$ symmetry is broken automatically by the phase coherence in the superconducting state. If symmetries other than $U(1)$ are kept, the superconducting state becomes conventional.

Superconductors that yield due to breaking additional symmetries besides the $U(1)$ symmetry [14] are unconventional superconductors. Thus, unconventional superconductors can be classified by determining the symmetry properties of the order parameter besides gauge symmetry.

By using the parity state in space of Cooper pairs, we can classify superconductors as those with even (spin singlet) and odd (spin-triplet) parities. The discovery of superconductors without inversion symmetry (noncentrosymmetric superconductors) such as $CePt_3Si$, UIr, $CeRhSi_3$, and $CeIrSi_3$ [18–21] has opened a new field to the theoretical study of superconductivity. The induced Rashba-type spin-orbit coupling due to absence of inversion symmetry leads to the superconducting state with mixed parity [22–25].

Motivated by new symmetry classes that appear in disordered superconductors, they have also gained great attention for theoretical studies [26–28]. In these systems, the single particle momentum is no longer a good quantum number and plain-wave eigenfunctions with momentum k should be replaced by position-dependent functions and pairing is between time-reversed states.

To find these functions, the generalized Hartree-Fock equations including the pairing potential of the superconducting state can be used, which leads to Bogoliubov-de Gennes (BdG) equations. The BdG equations are mostly used to describe the behavior of elementary excitation or quasiparticles that are created by Cooper pair breaking in the superconductors. Simultaneously, the properties and classification of the dirty superconductor can be determined by the BdG equations, where pairing symmetry is reflected.

When we apply magnetic field on the superconductor, depending on the strength of the magnetic field, the spin-polarized (strong magnetic field) or BCS states (weak magnetic field) will be favored. In the intermediate region of magnetic field, the Cooper pairs with opposite spins and net momentum are formed and Fulde-Ferrell-Larkin-Ovchinnikov (FFLO) state is realized [29, 30]. In this state, both translational and rotational symmetries are broken. This state is attractive for particle and nuclear physicists, because it is realized in high-density quark matter, nuclear matter, and cold fermionic atom systems [31]. In addition to FFLO states, there are other types of fascinating superconductors with multiple broken symmetry, such as ferromagnetic superconductors. According to the BCS theory, Cooper pairs in s-wave superconductors could be destroyed by an exchange field, which is known as paramagnetic effect, and it can be observed within a specific temperature interval with the maximum paramagnetic Meissner temperature.

Although, ferromagnetism and superconductivity have opposed order much attention has been paid due to nontrivial phenomena which predicted theoretically or found experimentally [32–34]. Ferromagnetic superconductors are expected to have triplet pairings since triplet pairings and ferromagnetism can coexist. So far, several ferromagnetic superconductors in bulk materials, for example, UGe 2 [35], ZrZn 2 [36], and URhGe [37], have been identified.

As mentioned earlier, symmetry is a cornerstone in the theory of superconductivity, and the properties of the Cooper pairs, that is, the basic constituents of superconductors, are determined by symmetry. Mathematically, correlation of two electrons to make Cooper pair depends on the position, spin, and time coordinate of these electrons. In BCS theory, the time coordinate is usually ignored but the symmetry property of a paired state allows for the interesting possibility that the two electrons are not correlated at equal times and that they are instead correlated as the time separation grows. In fact, this is realized if the correlation function is odd in time. This novel type of superconducting correlation that is odd in relative time or frequency is known as odd-frequency pairing and it has attracted a lot of attention. The pairing mechanism of odd-frequency superconductors had been theoretically introduced by Berezinskii in the context of superfluid ^{3}He [38]. This pairing mechanism was also predicted in ferromagnet/conventional superconductor junctions due to the breakdown of symmetry in spin space [33, 34]. The experimental results show that odd-frequency pairing exists near the interface in normal metal/superconductor junctions due to the violation of translational symmetry [39]. Consequently, the symmetry breaking more than U(1) is important for existence odd-frequency pairing. The emerging field of complexity in strongly correlated electronic systems may be connected to multiple symmetry breaking systems [40].

Quite analogously with noncentrosymmetric superconductors without inversion symmetry, where even- and odd-parity components are mixed, a mixing of even- and odd-frequency components should occur in principle, when the time reversal symmetry is broken, which affects many superconducting properties [41]. For example, the mixing of the conventional *s*-wave singlet with the odd-frequency triplet component in the presence of uniform magnetic fields, which may also be related to superconductivity coexisting with ferromagnetism, was proposed [42].

In last few years, technological advancement allowed to build superconductors of mesoscopic size like single electron transistors, quantum wires, quantum dots, quantum Hall systems, normal metal-superconductor-ferromagnet hybrid structures, magnetic multilayer systems, carbon nanotubes, graphene, small metallic nanoparticles, and nanomechanical systems, which are considered for both experimental and theoretical investigations [43–45]. In comparing with natural systems, artificial systems are of more interest for investigations because their transport properties can be measured in a more controllable way.

The superconducting properties of mesoscopic superconductors, in which one or more dimensions are comparable with the coherence length, differ remarkably from those of bulk superconductors and their properties such as the critical temperature and the critical fields are affected by the size of the sample. On the other hand, thermal and quantum fluctuations, which are strongly enhanced in these systems, cause dissipation and prevent the formation of the long-range-ordered state.

The study of emergent phases, phase competition, phase transitions, interactions between particles, and dimensional confinement in superconductors, which are of basic interest in the fundamental physics of magnetism, superconductivity, magnetoelectric and spin-dependent transport, is the central goal of the researchers. Their activities may produce new intuition on several related and persistent open questions in condensed matter physics and the results of these studies may also have practical implications for potential applications in spintronics and superconducting micro- or nano-electronics devices. The surface states of topological insulators have attracted strong interest in the past few years. The unique properties of the surface states provide a favorable field to explore emergent phenomenon, such as Majorana fermion-like excitations and quantum anomalous Hall effect. They are also considered attractive candidates for spintronic devices with pure spin current that can be controlled by electric field.

Author details

Heshmatollah Yavari
Department of Physics, University of Isfahan, Isfahan, Iran

*Address all correspondence to: heshmatollahyavari@gmail.com

References

[1] Onnes HK. The discovery of superconductivity. Communication Physics Laboratories. 1911;**12**:120

[2] Meissner W, Ochsenfeld R. Ein neuer Eekt bei Eintritt der Supraleit ahigkeit. Naturwissenschaften. 1933;**21**:787

[3] London F, London H. Supraleitung und diamagnetismus. Physica. 1935;**2**:341

[4] Ginzburg VL, Landau LD. On the theory of superconductivity. Zhurnal Eksperimental'noi i Teoreticheskoi Fiziki. 1950;**20**:1064

[5] Bardeen J, Cooper LN, Schrieffer JR. Theory of superconductivity. Physics Review. 1957;**108**:1175

[6] Bednorz JG, Müller KA. Possible highT_c superconductivity in the Ba–La–Cu–O system. Zeitschrift für Physik B. 1986;**64**:189

[7] Wu MK, Ashburn JR, Torng CJ, Hor PH, Meng RL, Gao L, et al. Superconductivity at 93 K in a new mixed-phase Y-Ba-Cu-O compound system at ambient pressure. Physical Review Letters. 1987;**58**:908

[8] Haeda H, Tanaka Y, Fukutami M, Asano T. A New High-Tc Oxide Superconductor without a Rare Earth Element. Japanese Journal of Applied Physics. 1988;**27**:L209

[9] Sheng ZZ, Hermann AM. Bulk superconductivity at 120 K in the Tl–Ca/Ba–Cu–O system. Nature. 1988;**332**:55

[10] Steglich F, Aarts J, Bredl CD, Lieke W, Meschede D, Franz W, et al. Superconductivity in the presence of strong Pauli paramagnetism: $CeCu_2Si_2$. Physical Review Letters. 1979;**43**(25):1892-1896

[11] Stewart GR. Heavy-fermion systems. Reviews of Modern Physics. 1984;**56**(4):755-787

[12] Jerome D, Mazaud A, Ribault M, Bechgaad K. Superconductivity in a synthetic organic conductor (TMTSF)2PF 6. Journal de Physique Letttres. 1980;**41**:L95

[13] Tanaka Y, Kashiwaya S. Theory of Tunneling Spectroscopy of d-Wave Superconductors. Physical Review Letters. 1995;**74**:3451

[14] Sigrist M, Ueda K. Phenomenological theory of unconventional superconductivity. Reviews of Modern Physics. 1991;**63**:239

[15] Mineev VP, Samokhin KV. Introduction to Unconventional Superconductivity. New York: Gordon and Breach Science; 1999

[16] Tsuei CC, Kirtley JR. Pairing symmetry in cuprate superconductors. Reviews of Modern Physics. 2000;**72**:969

[17] Damascelli A, Hussain Z, Shen Z-X. Angle-resolved photoemission studies of the cuprate superconductors. Reviews of Modern Physics. 2003;**75**:473

[18] Bauer E, Hilscher G, Michor H, Paul C, Scheidt EW, Grib-anov A, et al. Heavy Fermion Superconductivity and Magnetic Order in Noncentrosymmetric $CePt_3Si$. Physical Review Letters. 2004;**92**:027003

[19] Akazawa T, Hidaka H, Kotegawa H, Kobayashi T, Fujiwara T, Yamamoto E, et al. Pressure-induced Superconductivity in UIr. Journal of the Physical Society of Japan. 2004;**73**:3129

[20] Kimura N, Ito K, Saitoh K, Umeda Y, Aoki H, Terashima T. Pressure-Induced Superconductivity in

Noncentrosymmetric Heavy-Fermion $CeRhSi_3$. Physical Review Letters. 2005;**95**:247004

[21] Sugitani I, Okuda Y, Shishido H, Yamada T, Thamizhavel A, Yamamoto E, et al. Pressure-Induced Heavy-Fermion Superconductivity in Antiferromagnet $CeIrSi_3$ without Inversion Symmetry. Journal of the Physical Society of Japan. 2006;**75**:043703

[22] Rashba EI. Svoistva poluprovodnikov s petlei ekstremumov. I. Tsiklotronnyi i kombinirovannyi rezonans v magnitnom pole, perpendikulyarnom ploskosti petli Fizika Tverdogo Tela. 1960(6):1224-1238 [E.I. Rashba, Properties of semiconductors with an extremum loop .1. Cyclotron and combinational resonance in a magnetic field perpendicular to the plane of the loop. Soviet Physics Solid State. 1960;**2**:1109-1122]

[23] Edelstein VM. Characteristics of the Cooper pairing in two-dimensional noncentrosymmetric electron systems. Soviet Physics—JETP. 1989;**68**:1244

[24] Edelstein VM. Magnetoelectric Effect in Polar Superconductors. Physical Review Letters. 1995;**75**:2004

[25] Edelstein VM. The Ginzburg - Landau equation for superconductors of polar symmetry. Journal of Physics. Condensed Matter. 1996;**8**:339

[26] Altland A, Zirnbauer MR. Nonstandard symmetry classes in mesoscopic normal-superconducting hybrid structures. Physical Review B. 1997;**55**:1142

[27] Wigner EP. Proceedings of the On the statistical distribution of the widths and spacings of nuclear resonance levels. Cambridge Philosophical Society. 1951;**47**:790

[28] Wigner EP. On the Distribution of the Roots of Certain Symmetric Matrices. Annals of Mathematics. 1958;**67**:325

[29] Fulde P, Ferrel A. Superconductivity in a Strong Spin-Exchange Field. Physics Review. 1964;**135**:A550

[30] Larkin A, Ovchinnikov Y. Nonuniform state of superconductors. Soviet Physics—JETP. 1965;**20**:762

[31] Casalbuoni R, Narduli G. Inhomogeneous superconductivity in condensed matter and QCD. Reviews of Modern Physics. 2004;**76**:263

[32] Buzdin AI. Proximity effects in superconductor-ferromagnet heterostructures. Reviews of Modern Physics. 2005;**77**:935

[33] Bergeret FS, Volkov AF, Efetov KB. Long-Range Proximity Effects in Superconductor-Ferromagnet Structures. Physical Review Letters. 2001;**86**:4096

[34] Bergeret FS, Volkov AF, Efetov KB. Odd triplet superconductivity and related phenomena in superconductor-ferromagnet structures. Reviews of Modern Physics. 2005;**77**:1321

[35] Saxena SS, Agarwal P, Ahilan K, Grosche FM, Hasel-wimmer RKW, Steiner MJ, et al. Superconductivity on the border of itinerant-electron ferromagnetism in UGe_2. Nature. 2000;**406**:587

[36] Pfleiderer C, Uhlarz M, Hayden SM, Vollmer R, Lohneysen HV, Bernhoeft NR, et al. Erratum: Coexistence of superconductivity and ferromagnetism in the d-band metal $ZrZn_2$. Nature. 2001;**412**:58

[37] Aoki D, Huxley A, Ressouche E, Braithwaite D, Flouquet J, Brison J, et al. Coexistence of superconductivity and ferromagnetism in URhGe. Nature. 2001;**413**:613

[38] Berezinskii VL. Novaya model' anizotropnoi fazy sverkhtekuchego He^3 Pis'ma. New model of the anisotropic phase of superfluid He^3 Zhurnal Eksperimental'noi i Teoreticheskoi Fiziki. 1974;**20**:628. [JETP Lett. 20, 287 (1974)]

[39] Tanaka Y, Golubov AA, Kashiwaya S, Ueda M. Anomalous Josephson Effect between Even- and Odd-Frequency Superconductors. Physical Review Letters. 2007;**99**:037005

[40] Dagotto E. Complexity in strongly correlated electronic systems. Science. 2005;**309**:257

[41] Bauer E, Sigrist M, editors. Non-Centrosymmetric Superconductors. Berlin Heidelberg: Springer-Verlag; 2012

[42] Matsumoto M, Koga M, Kusunose H. Coexistence of Even- and Odd-Frequency Superconductivities Under Broken Time-Reversal Symmetry. Journal of the Physical Society of Japan. 2012;**81033702**

[43] Glattli C, Sanquer M, Trân Thanh Van J, editors. Mesoscopic Scales. Moriond Condensed Matter Physics Series. Editions Frontieres; 1999

[44] Awschalom DD, Loss D, Samarth N, editors. Semiconductor Spintronics and Quantum Computation. Berlin: Springer; 2002

[45] Nazarov YV, editor. Quantum Noise in Mesoscopic Physics. Dordrecht: Kluwer; 2003

Chapter 2

Superconductivity and Microwaves

Rafael Zamorano Ulloa

Abstract

Superconductivity conforms to a quantum, thermal, and electrodynamic set of physical phenomena of great interest by themselves. They have, also, the potential to be one clean energy source that technology is looking for. Superconductors do not allow static magnetic fields to penetrate them below a critical field, that is, Meissner effect. However, microwave magnetic fields do penetrate them already, and their energy is readily absorbed by the superconductor. High-temperature, perovskite superconductors do absorb microwave energy the most due to the presence of unpaired electron spins, fluxoid dynamics, and quasiparticle motion. We describe the fundamental physics of the interaction of the superconductors with microwaves. Experimental techniques to measure microwave absorption are presented. Experimental setups for absorption of energy are described in terms of the central quantity, Q. The measurements are analyzed in terms of irreversible energy exchange processes. The knowledge gained can inform the design of superconducting devices operating in microwave environments.

Keywords: superconductivity, microwaves, absorption, resonant cavity, superconductors type I, superconductors type II, HTSC

1. Introduction

Superconductivity (SC) conforms to an electrodynamic, quantum, and thermal set of physical phenomena of great interest by themselves [1–4]. A great amount of new fundamental physics has been extracted experimentally and theoretically since its discovery in 1911 by Onnes [4–17]. They have, also, the potential to be one clean energy source that technology is looking for. Superconducting solenoids could store clean magnetic energy to provide electricity to an entire house or set of houses for weeks, months, or years (**Figure 1A**). The actual electricity supply is shown in **Figure 1B**, SC solenoids installed at the house or set of houses, or factories with a simple system as shown in **Figure 1B** could provide an equivalent electrical energy to cover the domestic and industrial needs. The challenge is to charge the solenoid with a simple mechanism and control the discharge of very little amounts of energy, for their use, at a time. Leakage should be avoided at all times [18–20].

One of the most popular potential applications of superconductivity (SC) is the levitated train that can carry hundreds of people through hundreds of kilometers from one city to another at speeds in excess of 550 kms/h. in a very safe trip. A prototype is shown in **Figure 1D** and follows the same principle of the levitated magnet (**Figure 1C**). Several countries are developing real-scale prototypes, and engineering problems are being faced one by one [21, 22]. The above applications rely on just

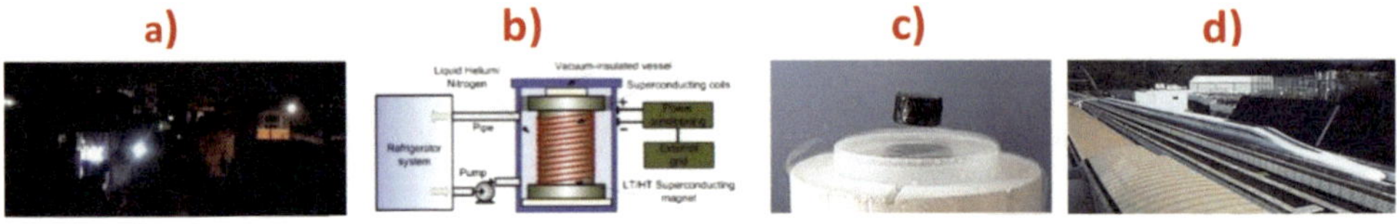

Figure 1.
(A) A house or a set of houses electrically powered conventionally could be powered with a superconducting reservoir of energy, kind of a cylinder as if it were a dynamo or a car alternator, (B) a sealed SC solenoid storing thousands of KWTTS of usable electrical energy [18–20], (C) a levitating magnet on top of a SC disc, and (D) a levitated train running on a bed of air because it has been levitated as the magnet at the left [21, 22].

two properties of superconductors, namely, the Meissner effect and the transport of electrical current with no resistance. The problem of interfacing with the conventional conductors that connect to machines, motors, etc. is not considered here.

2. Other technological uses of superconductors include electromagnetic cavities and microwaves

Pippard in 1947 reported an article [7]: "the first in a series in which the behaviour of the electrical impedance of metals at low temperatures and very high frequencies will be considered from experimental and theoretical standpoints. The technique of resonator measurements at 1200 Mcyc./s. is described in detail, and experimental curves are given showing the variation with temperature of the r. f. resistivities of superconducting tin and mercury. In contrast to the behaviour of superconductors in static fields, a finite resistance is present at all temperatures, tending as the absolute zero is approached to a very low value, which is probably zero for mercury but not for tin. The experimental results are in good agreement with London's measurements on tin by a different method."

Since then, spectacular discoveries and applications of superconductivity have been at work for decades in different areas of engineering, high-energy particle physics and microwave technology, communications, and scientific and medical (nuclear magnetic resonance) equipment [23–27].

At radio frequency and microwave range, SC has found applications in areas such as satellite communications, particle accelerators, microwave technology applied to precision instruments, and metrology and qubit formation [26, 27]. In particular, high-energy physics laboratories use both, SC solenoids and electromagnetic cavities [28, 29]. In one case, the production of dc strong magnetic fields, in the other high electric field at GHz is required.

Superconducting niobium cavities, the heart of linear accelerators: more specifically, to discover, experimentally, new elementary particles, or to investigate what are they made of, it is necessary to smash them against other high-energy particles. They are made to collide head-on at the highest attainable energies possible, ~500 Gev. The electron-positron colliders [30] need to work with particles at very high kinetic energy; hence their velocity must be as high as possible. To reach such high velocities, it is mandatory to accelerate the charged particles with extremely high electric fields, so that the force $F = qE$ acts and $F = ma$ makes its work accelerating the charged beam of particles (**Figure 2A**). The most efficient way (cost and physics) to accelerate electron and positron beams is to produce very high electric, E, fields, inside a connected series of SC niobium resonant cavities and to expose serially the charged particle beams to the electric field and let Newton's second Law to operate for as long as possible, as shown inside the red oval in **Figure 2A**. This part is the heart of the accelerating process and is supported

DOI: http://dx.doi.org/10.5772/intechopen.86239

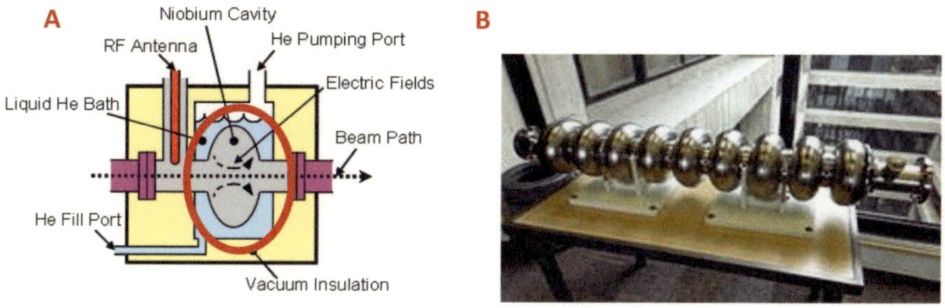

Figure 2.
A simplified diagram of a superconducting microwave cavity in a helium bath with microwave coupling and a passing particle beam. (A) the red oval encloses the heart of the accelerating process, (B) a niobium-based 1.3 GHz nine-cell superconducting microwave to be used at the main linac of the international linear collider.

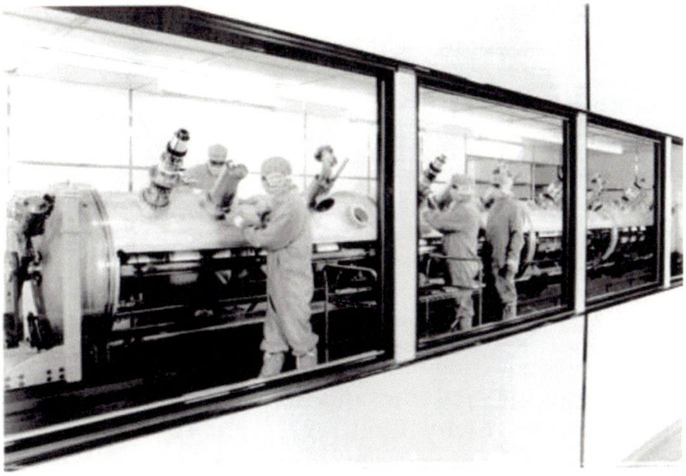

Figure 3.
Working to protect the Q of superconducting cavities. Four-cavity module for LEP being equipped with HOM and power couplers in a class 100 clean room [29]. Any impurity at the level 10^{-9} would produce heat spots, while the cavities operate to deliver, of the order of 500 Gev energy to the passing electrons.

by other fundamental parts as is the liquid helium bath to keep the temperature at 4.2 K, below the critical temperature for niobium to become SC.

Superconducting microwave cavities can sustain very concentrated patterns of electromagnetic fields with very, very small losses at the walls of the cavity; the surface impedance, Zs, is extremely low; the quality factor (Q) of a SC cavity is very high, $>10^5$; and the losses, proportional to 1/Q, are very small. Recalling the definition of Q = (ωo) stored electrodynamic energy/power loss [31, 32], where ωo is the operating frequency (~GHz), we see that the losses are just one part in 10^5 (or less) of the electrodynamic energy inside the cavity. The photo in **Figure 3** shows the work on a tandem of four superconducting cavities to mount detectors, and so on, inside a 100 clean room. The conversion of energy from the electromagnetic to the charged particle acceleration is the highest technologically possible at present [29, 30]. Acceleration of charged particles inside conventional conducting (copper) cavities presents much more losses at the cavity walls, Qcu ≪ Qsc and ZsCu ≫ZsSC, and the heat produced can melt the cavity. Accelerating charged particles in open space can disperse, and they cannot, by far, reach the concentrated fields and intensities that can be tailored inside SC cavities as the ones shown in **Figure 2**. The heart of the accelerating process is inside the superconducting cavity and inside it, in the red oval. The central piece in a linear particle accelerator is a tandem of these SC accelerating microwave cavities (**Figure 2B**).

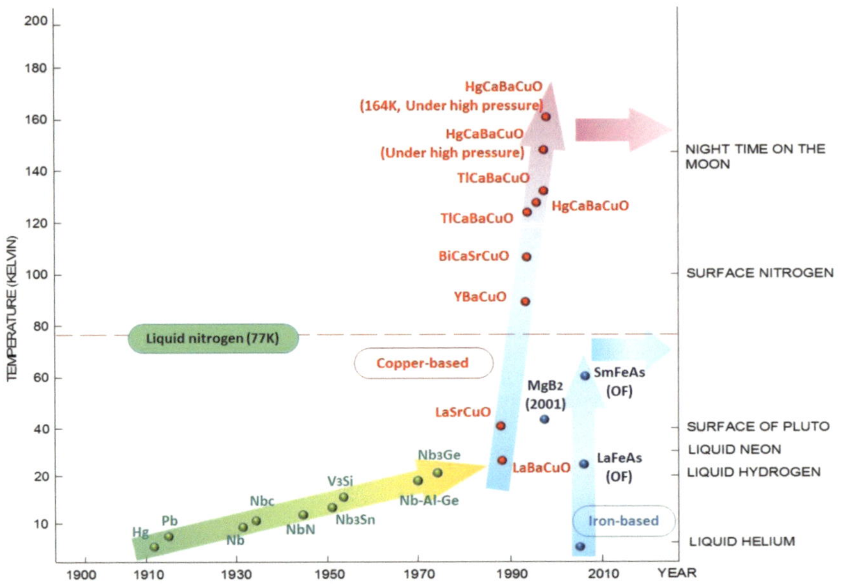

Figure 4.
Compilation of superconducting materials' families, starting with mercury in 1911 and stretching to the last discovery in 2018 [41]. The critical temperatures go from less than 1 K up to ≈ 260 K for LaH_{10} under 185 GPa (not shown). The comparison with other well-known temperatures in the world, placed on the vertical axes at the right, is illustrative.

Other salient superconducting microwave cavity applications and uses are (A) the measurement of fundamental constants through the electrodynamic response of a superconductor inside a cavity resonator [33] and (B) the measurement of superconductor losses at microwave frequencies dependent on applied magnetic fields and/or on temperature [7–8, 34–37]. The actual and the potential applications, just briefly mentioned here, are a sample of many propositions that are treated at length in Refs. [38–40]. The world of applications is so vast and promising that pushes the field of discovering new superconducting materials at higher and higher temperatures. **Figure 4** shows a compilation of several families of superconductors discovered at different times from 1911 up to the present (≈110 years).

SC comprises a large family of materials with a wide range of complex properties, behaviors, and types. They are classified in different ways: conventional, non-conventional, type I, type II, gapless, s-wave, d-wave, and so forth. To understand more precisely the interactions of superconductors with microwaves, we, next, highlight seven of the most fundamental characteristics of superconductors.

3. Eight of the most fundamental characteristics of superconductors

A. Superconducting critical temperature, Tc: a superconductor becomes so below a critical temperature, Tc. The superconducting state is ordered, and so, it is destroyed by thermal agitation energy, k_BT. But, as temperature is lowered, the thermal energy goes down and the atomic and molecular giggling becomes quitter, and the superconducting state produced by the pairing of electrons is not so disturbed by the thermal energy k_BT below k_BTc. We could say that the two energies, the superconducting, ordering energy, Δ, compete with k_BT. In a broad picture, when $k_BT >$ ordering energy (Δ), the material ceases to be superconductor, and when $k_BT <$ (Δ), the material becomes superconducting.

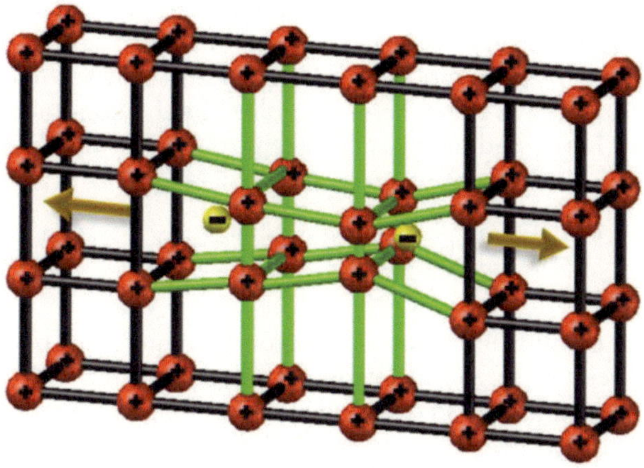

Figure 5.
A lattice of atoms is distorted by the passage motion of two electrons, possibly tracing time reversal trajectories [42]. The center of mass does not move and the pair is in a singlet state. The passing electrons attract the positive-charged nuclei of the lattice, causing a slight ripple in its wake.

B. The formation of Cooper pairs. The ordering consists of the forming of electron-bound pairs, called Cooper pairs, by means of an attractive interaction mediated by the lattice vibrations (phonons) that overcomes the Coulomb repulsion that experience these electrons. **Figure 5** depicts in an exaggerated manner the lattice distortion when the couple of electrons move through this part of the material. Through the vibrations of the lattice, these two electrons form a bound system with zero linear momentum p and zero spin, a singlet Cooper pair.

C. The macroscopic quantum state $\psi(r, t)$, Cooper pair, coherence length, and penetration depth. These Cooper pairs are bosons (spin zero), and they all condensate in the lowest energy state, and all has exactly the same wave function, $\psi(r, t) = \psi_0 \exp(i\phi)$, where φ is now the macroscopic phase of the wave function. The whole macroscopic quantum state is represented by $\psi(r, t) = \psi_0 \exp(i\phi)$. The square of the wave function gives the probability of finding a Cooper pair, let say, at r*, at time t*, but such probability is exactly the same to find another Cooper pair at such r* and t* and another Cooper pair at r* at t*. If there are ns bosons in the superconductor, after normalization, we have $|\psi(r^*, t^*)|^2 = ns|\exp(i\phi)|^2 = ns$, the total number of Cooper pairs in the superconductor [1–3, 17]. The probability of finding a Cooper pair at r*, t* is the same as finding another Cooper pair at any other location at any other time inside the superconductor. $\psi(r, t)$ is also the complex order parameter first introduced by Ginsburg-Landau in their 1950s superconductivity theory [10] in which the superconducting electrons were envisioned as to enter (a second-order transition) or form a superfluid state, ψ, which is an ordered state.

The Ginzburg-Landau equations [43] predicted two new characteristic lengths in a superconductor. The first characteristic length was termed *coherence length*, ξ. For $T > T_c$ (normal phase), it is given by:

$$\xi = \sqrt{\frac{\hbar^2}{2m\,|\alpha|}} \qquad (1)$$

where α is a parameter in the GL Equations [43]. While for $T < T_c$ (superconducting phase), where it is more relevant, it is given by

$$\xi = \sqrt{\frac{\hbar^2}{4m\,|\alpha|}} \qquad (2)$$

It sets the exponential law according to which small perturbations of density of superconducting electrons recover their equilibrium value ψ_0. Thus this theory characterized all superconductors by two length scales. The second one is the penetration depth, λ. It was previously introduced by the London brothers in their *London theory* [6]. Expressed in terms of the parameters of Ginzburg-Landau model, it is.

$$\lambda = \sqrt{\frac{m}{4\mu_0 e^2 \psi_0^2}} \tag{3}$$

where ψ_0 is the equilibrium value of the order parameter in the absence of an electromagnetic field. The penetration depth sets the exponential law according to which an external magnetic field decays inside the superconductor. The ratio $\kappa = \lambda/\xi$ is known as the Ginzburg-Landau parameter. It has been proposed by Landau that *Type I superconductors* are those with $0 < \kappa < 1/\sqrt{2}$ and *Type II superconductors* those with $\kappa > 1/\sqrt{2}$.

D. Zero electrical resistance. Supercurrents, Js, of Cooper pairs. Superconducting conductivity, σs. No losses versus normal conductivity, σn. Losses and heating. When these Cooper pairs move in the material, they do so in phase with the lattice vibrations, and no scattering of electron pairs occurs and no electrical resistance develops, and very importantly, no dissipation of energy develops due to this charge transport (see blue portion of the temperature axis and coupled blue dots (Cooper pairs) in **Figure 6A** and **B**). In contrast, when single conduction electrons move in a normal metal or in YBaCuO perovskites, at temperatures above Tc (red portion and red single dots in **Figure 6A** and **B**), they scatter many times from the nuclei, imperfections, impurities. They lose energy in each one of these processes, giving rise to the well-known and measurable electric resistance and/or normal conductivity, σn. The material develops Joule heat, and for high currents (hundreds to thousands of amperes), the metal can even melt. But, in the SC state, the electrical resistance is zero, the whole conductivity is σs, and thousands of amperes can flow without losses nor Joule heating.

E. Meissner effect: magnetic fields get expelled. When a superconductor is placed in an external magnetic field, or a material becomes SC in the presence of such external magnetic field, the SC material expels the magnetic field from its interior, as shown in **Figure 7A**. Irrespective of the state and history of the material

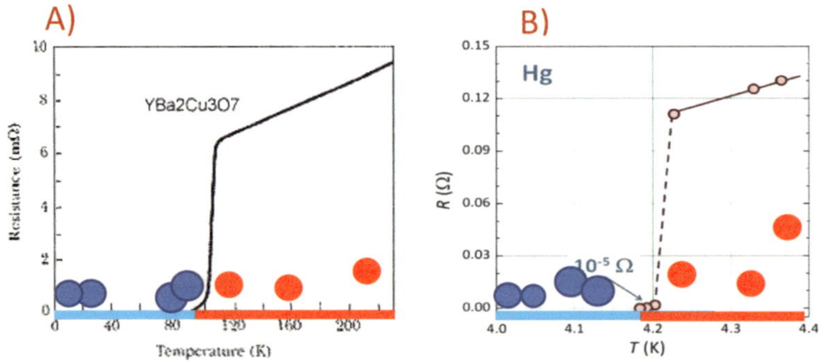

Figure 6.
For a SC material, above the critical temperature, Tc, the conductivity is "normal". If the material is a classic conductor, then it shows normal conductivity, σn, above Tc. (A) if the material is a cuprate (v.gr. A YBaCuO perovskite superconductor), above Tc, its conductivity is "normal" but very low. Once the temperature is decreased below Tc, the material undergoes a phase transition to the SC state and the resistance goes to zero, and the conductivity is σs. (B) we show the historic case of this measurement for Hg, made by Onnes in 1911, when the superconducting state was discovered.

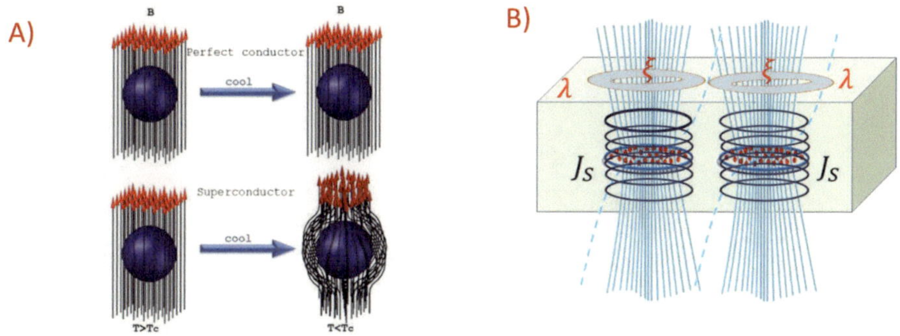

Figure 7.
(A) 3D effect of the presence of a superconductor in a region of magnetic field. The lines of magnetic field get expelled from the interior of the SC body. In contrast, a perfect conductor (defined as a material with conductivity going to infinity and resistivity going to zero) will allow the magnetic field lines to penetrate it, at any temperature (there is no transition to anything here). Its behavior is the same at all temperatures, (B) the flux lines penetrate and cross the whole material in "ramilletes" or bundles of very small radii ξ when the SC is in the mixed state, $Hc_1 < H < Hc_2$.

prior to the superconducting state, once in it, all magnetic lines of the external magnetic field get EXPELLED from its interior. This experimental result was discovered by Meissner and Ochsenfeld in 1933 and is known as the Meissner effect. By comparison, a perfect conductor is penetrated by the magnetic field lines irrespective of the temperature. In great contrast, the superconducting material expels the field lines while in the superconducting state, T < Tc. Above Tc, the material behaves as a normal conductor, and so, the magnetic field lines go through it.

F. Type I superconductor, one critical magnetic field. Type II superconductors. Two critical magnetic fields, Hc1 and Hc2. The mixed state. The SC that shows the behavior shown in **Figure 7A** up to a field Hc, before the SC state breaks down into the normal state, is said to be of the first kind or type I SC. But not all superconductors are of the first kind. A second kind of SC materials presents the expulsion of magnetic field lines' behavior up to a critical field, Hc1, which is much smaller than the critical field that withstand the superconductors of the first kind, Hc1 ≪ Hc. For fields above Hc1, the SC of the second kind admits the entrance of magnetic field lines but in an ordered and quantized manner. These flux lines penetrate and cross the whole material in "ramilletes" or bundles of very small radii ξ as shown in **Figure 7B**. These lines define approximate cylinders that can be bent and twisted. Inside these flux cylinders or fluxoids, the material is in the normal state, and outside the fluxoids, the material is superconducting. This configuration is allowed in the mixed state, $Hc_1 < H < Hc_2$. The cylinders are surrounded by supercurrents of Cooper pairs (gray discs in **Figure 6**) that limit the expansion of the radius of the cylinder. The flux is quantized, $\varphi n = n\,\varphi o = n(\hbar/2e)$.

Hence, coexistence of both normal and SC states is allowed in the mixed state. Inside these fluxoids, ξ, there are no Cooper pairs, but only normal electrons that if moved, they will do it in the normal way, σn, dissipating energy because inside the vortices, the medium is normal-viscous for the motion of the electrons. If alternating electromagnetic fields are applied to such superconductor in the mixed state, the Lorentz force on the normal electrons will force them into periodic damped motion; in addition, the inside of these vortices will radiate since it contains electric charges in accelerated motion [16, 31]. If microwaves are applied, again, the electrons are subjected to a Lorentz force that will make the normal electrons to move with friction and would be accelerated and radiate microwaves. This result was established more than 70 years ago [7, 8].

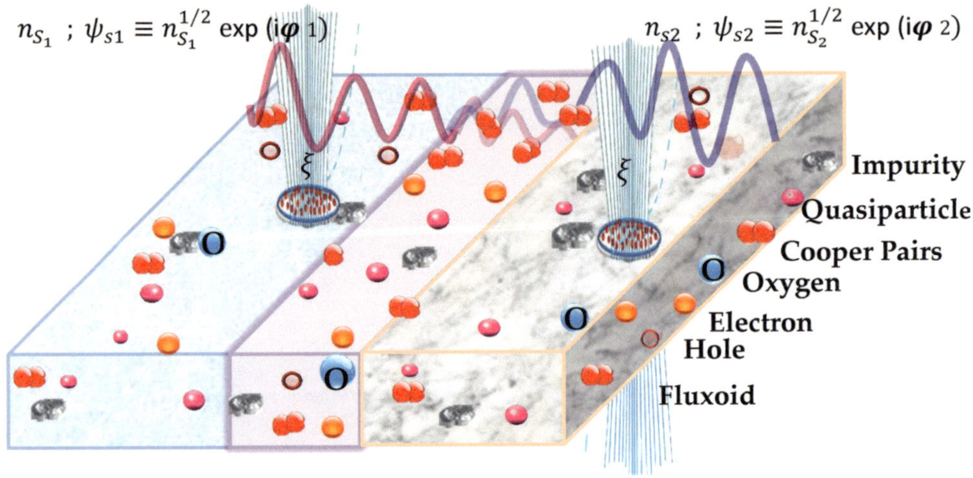

Figure 8.
Two superconductors in proximity separated by a layer of insulator, or conducting material, produce Josephson effects. The wave functions of each superconductor go into the isolator or conducting material (yellow) in the middle of the two superconductors, and the probability to find Cooper pairs in here is different from zero. As a result, transport of Cooper pairs appears from one superconductor to the other through the barrier [15, 16, 46].

G. When two superconductors, SC1 and SC2, are put in proximity, the non-intuitive Josephson effects emerge. When the surfaces of SC1 and SC2 are a few microns, or tenths, or hundredths of microns apart, several non-intuitive quantum phenomena happen. Josephson discovered them theoretically in 1962, and they are known as Josephson effects, for which he received the Nobel Prize in 1973. The main effects are J1 at the junction of two superconductors separated by a thin insulator, without any voltage applied (**Figure 8**), Cooper pairs can tunnel the barrier and a supercurrent is established between both superconductors [15, 44], and the current is governed by the equation:

$$Jsc = Jc \sin(\varphi_2 - \varphi_1) \qquad (4)$$

where φ_1 and φ_2 are the corresponding phases of the superconducting wave functions of each superconductor and Jc is the maxim current that the structure can withhold without producing NEW phenomenology and depends on material properties. Given that φ_1 and φ_2 are constant, then in Eq. (4) Jsc is constant and is usually termed the dc Josephson effect. The equation was first derived fully under theoretical considerations, for the possibility of tunneling Cooper pairs from one superconductor to the other as shown in **Figure 8**. Few months later, it was corroborated experimentally by Anderson and Rowell [45].

The materials depicted in **Figure 8** can be small enough to be point-contact electrodes.

The second main effect is known as the ac Josephson effect and is realized when a constant voltage is applied to the junction. Such applied voltage, V, causes an oscillating supercurrent, J2e, to flow across the junction and is given by

$$J2e(\omega t) = Jc \cdot \cos(\omega t) \qquad (5)$$

Where $\omega = 2eV/\hbar = E/\hbar$ and $E = h\nu = 2eV$. This equation tells us that the application of a CONSTANT voltage across the junction produces that the cooper pairs flow through the junction, J2e, as a periodic function given by cos(ωt). This is an exact transformation of a direct voltage (macoscopic) to a periodic response of the superconducting electrons, in which they move with the quantized energy: $E = \hbar\omega$. The frequency of the oscillating current depends only on the voltage applied, the charge of the Cooper pair (2e), and Plank's constant. The relation between voltage and frequency of produced current is exact and has served as a very precise voltage standard. This effect is fully quantum macroscopic, first deduced by Josephson and soon afterwards was confirmed experimentally. The energy hν equals the energy change of a Cooper pair transferred across the junction. The voltage applied to a Josephson junction is typically on the order of a few microvolts. Thus, Josephson junctions (JJ) and superconducting quantum circuits (SQCs) are usually operating at frequencies in the microwave regime, from a few GHz up to THz [46, 47]. Hence, a JJ or a SQC becomes a microwave source when injected a few dc microvolts.

When, in general, the phases of the wave functions are functions of time, then the Josephson equations are

$$V(t) = \frac{\hbar}{2e}\frac{\partial\phi}{\partial t} \text{ (superconducting phase evolution equation)} \tag{6}$$

$$Jsc = Jc\sin(\varphi_2(t) - \varphi_1(t)) \text{ (Josephson or weak - link current - phase relation)} \tag{7}$$

The rate of change of the phase difference ϕ directly proportional to the voltage V across the junction through the relation $\partial\phi/\partial t = 2\ eV/\hbar$. From these two equations, it can be shown that the Josephson junction has an energy $-E_J \cos\phi$, where $E_J = \hbar c/(2e)$ is called the Josephson energy. Furthermore, the current-voltage relations reveal that the Josephson junction functions like a nonlinear inductance with $L = \hbar/(2ec \cos\phi)$. It is this nonlinearity of the Josephson junction that brings about the anharmonicity of JJs and SQCs. In a given SQC, we can thus select the two lowest energy levels from the non-equally spaced energy spectrum. These two levels form a quantum bit (qubit) for quantum-information processing. When an ac voltage V is applied to the two electrodes of the JJ, the supercurrent I is periodically modulated as $I = Ic \sin(\omega t + \phi)$ with the Josephson frequency $\nu = \omega/(2\pi) = 2\ eV/h$, and the energy hν equals the energy change of a Cooper pair transferred across the junction.

H. Low-temperature superconductors (LTS) and high-temperature superconductors (HTSC). The discovery of HTSC in 1986 by Bednorz and Müller [48] and the following higher critical temperature superconductor discovery by Wu et al. [49] in 1987 and the following discoveries of even higher critical temperature superconductors (up to appro. 150 K) marked a great contrast with the already known low-temperature superconductors (LTSC). **Figure 4** shows a graph of the critical temperature of many superconductors as a function of the year they were discovered. It is apparent in the graph the great number of superconductors already known at present. Many of them belong to families of compounds, as cuprates, the actinides, the pure SC elements, and so forth. The SC that are well described by the BCS theory are called conventional superconductors (cSC), and the ones that resist the BCS theory are called non-conventional superconductors (ncSC). It is generally believed that the HTSC cuprates are not conventional, because their energy gap is characterized as d-wave in contrast to the s-wave energy gap of conventional SC [50]. However, such belief has been challenged very recently [51] and the debate follows. The point is that we do not know fully the whole SC state of these HTSC cuprates.

4. Superconductors generate and respond to microwaves: the effect of microwave fields on superconducting tunneling in JJs and on SQCs

Beyond theoretical issues, given the great potential for their technological applications, the characterization of their properties is, and had been, a very intense activity worldwide for many years producing thousands and thousands of research reports. Here we concentrate on the interactions of microwaves with superconductors as single samples and as Josephson junctions (structures). We pay attention to systems where weak links are present. We mention here that the dissipation processes are plenty. It could be a type I or type II cSC, or an ncSC, or JJs and SQCs being bathed by microwaves.

Shapiro [47] introduced several types of JJs at X-band and K-band microwave cavities and determined experimentally, for the first time, the effects of microwaves of 9.3 GHz and 24.85 GHz on them. He found the, now well-known, Shapiro steps in the I-V characteristic curve. The effect of microwave power in modifying the I-V curve, especially the zero-voltage currents, is evident in the trace in **Figure 9**. The samples were mounted in a microwave cavity resonant, at low temperatures. All the observations were independent of sweep frequency and were also seen when dc was passed through the sample. The interval in voltage from one zero-slope region to the next is not always hv/2e; sometimes a step is missing so that the voltage interval is hv/e.

Shapiro states "Josephson has already discussed briefly the effect rf should have on the pair-tunneling supercurrent. He predicted the occurrence of zero-slope regions separated by hv/2e in the I-V characteristic in the presence of the rf field. This prediction was based on the frequency modulation by the rf of the ac supercurrents previously referred to. Our experiments have confirmed this prediction and represent indirect proof of the reality of Josephson's ac supercurrent."

If a Josephson junction (JJ) is microwave irradiated, it absorbs part of its energy (which can be used as a detection mechanism), and it excites periodic dynamics that again produce microwaves. These processes have been already in use

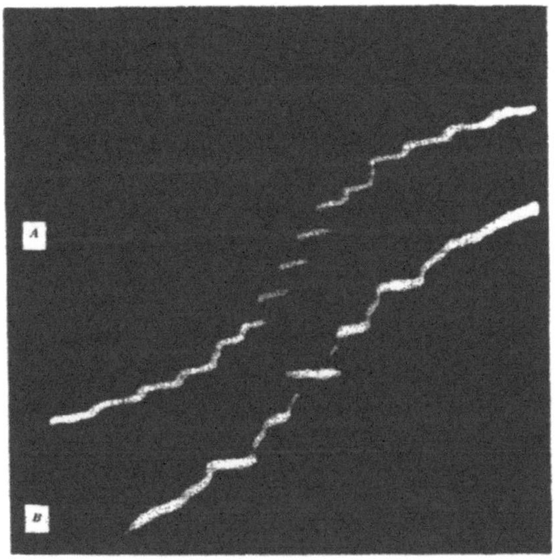

Figure 9.
Microwave power at 9.3 GHz (a) and 24.85 GHz (B) produces many zero-slope regions spaced at hv/2e or hv/e. for a, hv/e = 38.5 pV, and for B, 103 pV. For a, vertical scale is 58.8 pV/cm and horizontal scale is 67 nA/cm; for B, vertical scale is 50 pV/cm and horizontal scale is 50 pA/cm. Similar effects occur at X-band and at K-band [47].

to produce microwaves with superconducting junctions [46, 52]. The so-called weak link junctions are appropriate for the development of ultra-low-power microwave sources [52, 53].

Wertham and others [46, 52] have studied the nonlinear self-coupling of Josephson radiation in diverse superconducting tunnel junctions. Since the original prediction by Josephson of an ac supercurrent generated in a superconducting tunnel junction by a dc bias voltage, considerable work has developed in the electromagnetic radiation associated with the ac current. Detection of the produced radiation is complicated by the high degree of nonlinearity of the device, JJ, SQC-detector, together with the large amount of feedback intrinsic to its configuration, networks of weak links, or tandems of JJs. The structures radiate into the insulator (conductors) and the surrounding superconductor islands-grains, and the resulting electromagnetic fields will feed back to influence significantly the currents which generate them. The starting point would be Maxwell's equations for the electric and magnetic fields radiated from an oscillating charge distribution and then finding the radiation fields in the near zone [31]. Wertham shows detailed calculations and finds the ac tunneling current contains components at all harmonics of the fundamental frequency, ωo [52]. One would like to impose the requirement of very low loss in the radiating junctions, but that is inconsistent with the large amount of the external battery power being supplied by the dc current. The more power radiated, the more power supplied and more losses.

Very recently Tikhonov, Skvortsov, and Klapwijk have addressed the problem of nonequilibrium superconductivity in the presence of microwave irradiation [54]. Using contemporary analytical methods, they refine the old Eliashberg theory and generalize it to arbitrary temperature T and frequencies ω. Microwave radiation is shown to stimulate superconductivity in a bounded region in the (ω, T) plane. In particular, for $T < 0.47Tc$ and for $\hbar\omega > 3.3kBTc$, superconductivity is always suppressed by a weak ac driving. They also study the supercurrent in the presence of microwave irradiation and establish the criterion for the critical current enhancement. They interpret qualitatively in terms of the interplay between the kinetic ("stimulation" vs. "heating") and spectral ("depairing") effects of the microwaves [54].

Losses by heat, radiation, dissipative fluxoid dynamics, quasiparticle scattering, and other dissipating processes can limit considerably the SC performance as detector elements and in SC devices, or SQC, or as quantum bits (qubits). The microwave absorption profile of any given superconductor or Josephson junction is a fundamental measurement of performance, and it brings rich information on the SC electrons, the normal electrons, and the moving or stationary vortices and on their capacity to absorb and to produce microwaves in the range from GHz to THz.

Hence the response of the superconductors as bulk, as films, and as single arrays, or networks, of Josephson junctions to these electromagnetic excitations needs to be characterized experimentally. We focus on the measurement of losses in the laboratory using one of the most sensitive methods: perturbation cavity, the determination of the Q of the cavity, and its change due to material properties, or as a function of temperature.

5. Measurement of absorption of microwaves by superconductors. The experiment, the Q of electromagnetic cavities, the perturbation cavity method

One of the most sensitive measurement techniques to determine absorption of microwaves is the so-called perturbation cavity method. An electromagnetic resonant cavity made of very good conductor, or even superconductor (v. gr: niobium as for particle accelerators), reflects from its walls the electromagnetic fields inside it. Absorption by the walls is very low. Two are the fundamental characteristics of a cavity that have in its interior a small SC piece as a sample to measure how much microwave power it absorbs [7–8, 32, 34–37, 47]: the quality factor, Q, and the frequency shift of the resonance, Δf. These changes can be very small, and it is convenient to include a very selective, filtered, amplification stage, such as the technique of lock-in amplification. What is measured is the quality factor change, $\Delta Q = Q_S - Q_0$, from the quality factor, Q_0, from the empty cavity, and the frequency shift produced by inductive changes in the electromagnetic energy inside the cavity [35–37]. Remembering, the quality factor is defined as [31]:

$$Q = (\omega o)\ (\text{stored electrodynamic energy})/(\text{power loss}) \qquad (8)$$

Its inverse is 1/Q = [electromagnetic energy loss (dissipated) by the sample]/ ω_0(total electromagnetic energy inside the empty cavity).

Thus, effectively, the inverse of Q is a direct measurement of electromagnetic energy inside the cavity, other than the wall losses. Hence, Δ(1/Q) is a direct measurement of the electromagnetic absorption by the superconducting sample inside the cavity. Knowing the eight fundamental properties of superconductors briefly described above will allow an informed interpretation of the measurements and disentangle several contributions to the microwave losses. The type of superconductor, the state it is in, the level of impurities, the presence of JJs, and in particular, the presence of weak links would determine the superconducting processes taking place and which one of them can generate microwave absorption. A typical granular HTSC, or LTSC, conventional, or unconventional, superconductor with many processes pictorially represented (but not all that could operate) is shown in **Figure 10**. It should be noticed how crowded it looks. This indicates partly its complexity and how many things are going on inside the sample.

An electron paramagnetic resonance equipment where absorption of microwaves is measured very accurately is an adequate way to measure microwave absorption of SCs. Such measurement is realized by capturing the reflected

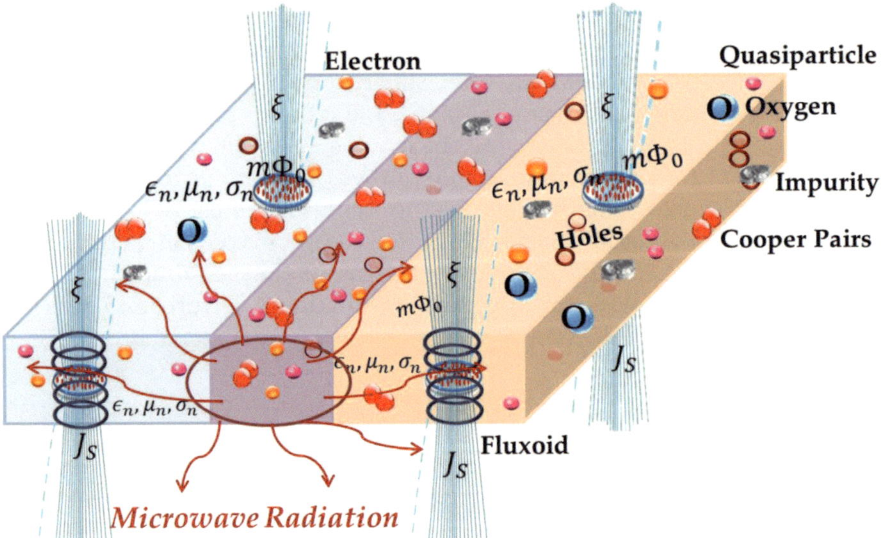

Figure 10.
A pictorial representation of a Josephson junction in which fluxoids move or are pinned and quasiparticles in the "middle" material (insulator, conductor, semiconductor) move and oscillate. There are impurities, and the cores of the vortices contain normal electrons, and the core behaves as a normal electromagnetic medium (ε, μ, σ). The Cooper pairs are present abundantly on both superconductors and tunnel frequently at the middle.

microwaves (not the absorbed microwaves) from the superconducting specimen and recombining (superposing) them with part of the incident waves (about 10% of these) within the microwave bridge, to reach the crystal detector. The crystal detector, in turn, generates a crystal current that is converted to voltage and then sent to the electronic filtering, amplification, and display units of the equipment. All these experimental stages are represented by blocks in part B of **Figure 11**. Typically, an oscilloscope trace shows the mode of the unloaded cavity and of the loaded cavity. From the display, the shifted, Δf, and width, ΔQ, and the reduced high with respect to the empty cavity modes, fo and Qo, are obtained (see red oval in **Figure 11B**). The shift and the broadening are, respectively, proportional to the inductive conductivity (more generally, lossless processes that alter the eigenresonance of the cavity), σ_1. The differences in the width, Δw, are proportional to the

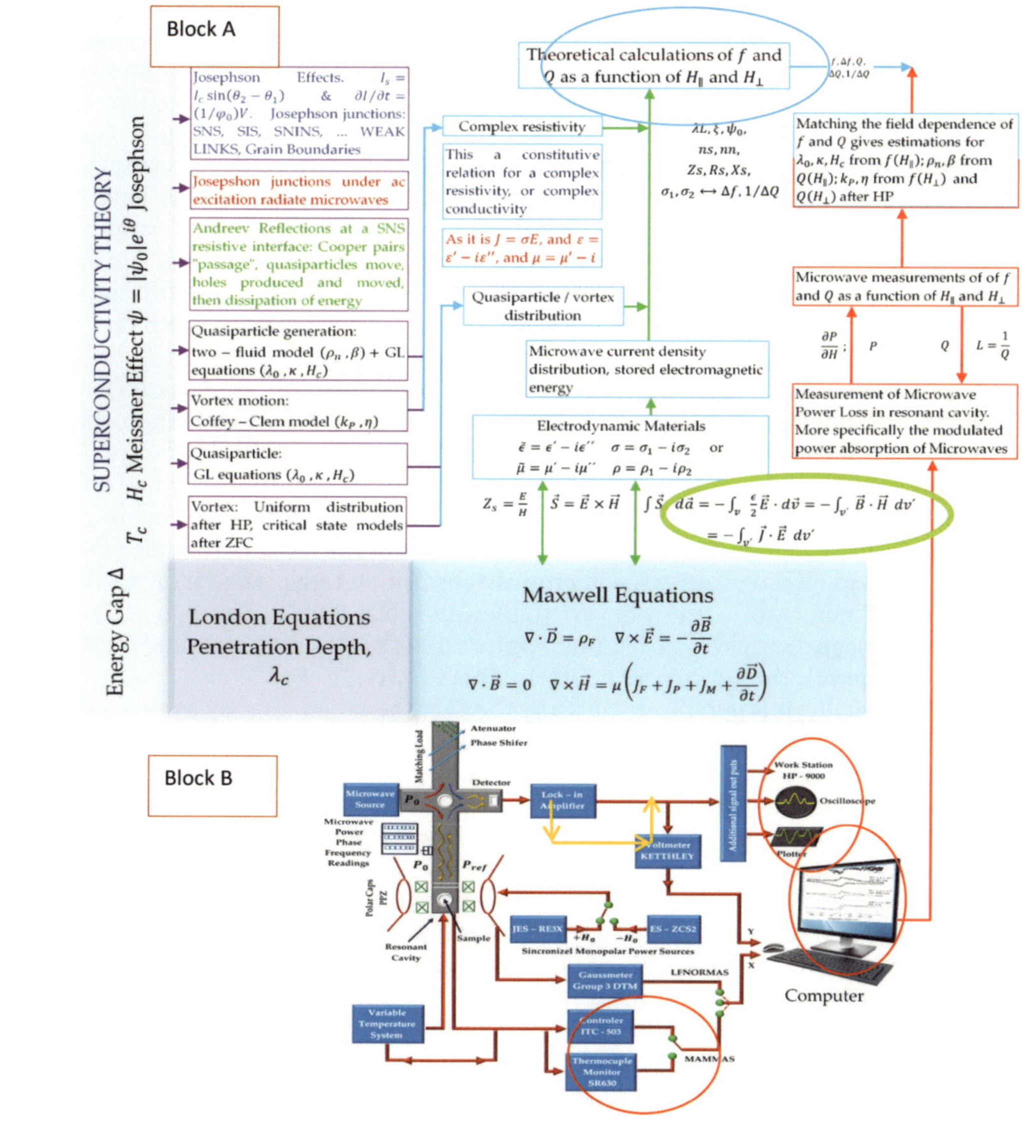

Figure 11.
The combined theory and experiment involved in microwave losses by superconductors. Theory includes Maxwell electrodynamics, with complex impedance and complex conductivity, and radiation terms inside resonant cavities. SC theory is large and complex, and few characteristics are represented in the SC blocks. The connection is the quality factor Q.

loss of electromagnetic energy inside the cavity and due to the SC sample inside it, and these losses are quantified by 1/Qs − 1/Qo. Now, how does the measurement of Q connects with the whole of the superconductivity theory and Maxwell equations?

6. The combined superconducting theory and experiment of microwave losses

The electrodynamics theory of E and B fields and losses, Q, in resonant cavities is well developed. Jackson [31], Csősz et al. [55], Kwon et al. [56], and other books [1, 2] have explicit equations for losses due to dielectric, metallic, or superconducting materials. Following Kwon et al., here we present in compact form (block diagram) (**Figure 11A**) a summary of the electrodynamics and losses of superconductors in resonant cavities. We CONNECT theoretical expressions with the experimental block diagram that measures Q (**Figure 11B**). The quantity that articulates both parts is the quality factor, Q, of the loaded cavity. On one hand, it can be arrived at from theory (BLUE OVAL), taking into account dissipative processes that develop in a superconducting sample or a Josephson junction. The SC theory represented by the leftmost blocks of the diagram feed the electromagnetic theory of Q with fundamental SC parameters, such as the coherence length, ξ; the London penetration depth, λ; critical temperatures; energy gap, Δ; values and critical fields, Hc, H1, and H2; their temperature and field behaviors; and so forth. On the other hand, we obtain Q by measurement (RED OVALS) with the equipment shown in **Figure 11B** or any other version of measurement with the perturbation cavity method.

Figure 11 shows two block diagrams connected by the quantity Q. Block A includes the fundamental theory of superconductivity and t, the main phenomena, effects, and fundamental parameters, and block B shows a commonly used experimental arrangement for measuring microwave losses by means of the technique of cavity perturbation. Here, we include a field modulation stage that can bring much more sensitivity to details of the recordings [35–37]. But, the modulation stage can be bypassed and record directly the microwave absorption (see orange arrows in **Figure 11**). If temperature is slowly varied, then one obtains Q(T) (from $P_{ABS}(T)$). Likewise, if magnetic field is varied (and excite a lot of vortex dynamics and quasiparticle dynamics), then one obtains Q(H) (from $P_{ABS}(H)$).

Experimentally, it is possible to measure the Q of the cavity directly as an output of the amplifier or directly from the detector used. If the output is being modulated, then the measurement is the derivative with respect to varying magnetic field, Ho < H1, or Hc1 < Ho < Hc2, if the superconductor is in the Meissner state or the SC sample is a type II superconductor in the mixed state. Our block diagram includes low-temperature equipment coupled to the resonant cavity, so the sample could be set at any temperature between 4.2 K and 77 K, or 300 K, or any desired temperature. Temperature can also be gradually changed in order to register Q(T). By field modulation, the derivative of Q(T) with respect to magnetic field can also be recorded. Much more structure is captured in this way. SC losses, as measured inside resonant cavities, continue to be vigorous due to the wealth of novel information that continues to be discovered.

Next, we take a SC system, granular, crystalline JJ, with strong or weak links, JJ networks, SQC, and so forth, that is bathed with microwaves coming from a particular direction; several portions of the sample will respond to microwaves differently. Take the JJ represented in **Figure 10**; quasiparticles and electrons and holes in the sandwiched (the "middle") material will execute dissipative accelerated motions due to the Lorentz force they experience. Vortices will also execute translational viscous-dissipative motions because, again, the Lorentz force (or Magnus force) drives them [1–3, 26]. Pinned vortices will execute damped oscillations, and this will absorb electromagnetic energy. The normal electrons inside the fluxoid cores will move along with the fluxoids themselves; their motion is, again, dissipative. If we, abstractly, enumerate the different loss processes we have mentioned and some others that could operate (as the electrons and the holes produced in the multiple Andreev reflections), then we can write: total loss = Lt = L1 + L2 + L 3 + L4 + ..., and 1/Qt = 1/Q1 + ...

We notice that several of these terms are of the same form with respect to the expression of the dissipated electromagnetic energy that enters into Poynting's theorem and in consequence into the Q expression:

The higher the impurities and/or defects, the higher the pinning centers, the higher the microwave losses due to both fluxoid dissipative dynamics, $(1/Q)_{impure}$. The higher the normal conductivity, σn, the lower the normal resistivity, and the larger the local mean free path that the normal charge carriers experience, and in consequence, the less charge scattering and less energy loss. Conversely, the lower the normal conductivity, the higher the normal resistivity and the higher the losses (1/Q). The higher the applied static field in the $Hc1 < Ho < Hc2$ range, the larger the number of fluxoids present or forming at the edge of the SC, $\Phi = n\Phi o$, the more elements executing dissipative dynamics and more overall microwave loss.

Varying temperature: given a superconductor, either HTCS or LTSC, the lower the temperature with respect to its critical temperature (Tc), let us say 22%, the more electrons form Cooper pairs and they will move around without dissipation. The electrons that remain normal are less and less as T departs from Tc; these electrons will produce less losses, $(1/Q)_T$, than the same material at, let us say, 77% of Tc. Hence, for identical materials, all other parameters fixed, we would expect $(1/Q)_{T = 22\%} < (1/Q)_{T = 77\%}$.

From the theory, the integrals of electromagnetic energy are related to the complex conductivity, $\sigma = \sigma n - i\sigma_{SC}$, and with the Q of the cavity. These integrals are a deduction of the electrodynamics of the superconductor based in Maxwell equations and the London equations. The superconducting parameters, ξ, λ, Δ, κ, σ_S, σ, are involved in these equations, and hence they determine also the microwave absorption. From the experimental side, we have represented by blocks the main constituents of the generation of microwaves, the resonant cavity, and the detection system. The quantity measured is the Q of the cavity. Contrasting the theoretical formulas with the measured absorption and knowing the type of SC sample and/or if it is Josephson junction, an assignment of the losses can be made.

As a manner of example, we take the case of magnetic field dependent microwave losses in superconducting niobium microstrip resonators reported by Kwon et al. [56]. They find that quasiparticle generation is the dominant loss mechanism for parallel magnetic fields. For perpendicular fields, the dominant loss mechanism is vortex motion or switches from quasiparticle generation to vortex motion, depending on the cooling procedures. They calculate the expected resonance frequency and the quality factor as a function of the magnetic field by modeling the complex resistivity. Key parameters characterizing microwave loss are estimated from comparisons of the observed and expected resonator properties. Based on these key parameters, they find that a niobium resonator whose thickness is similar to its penetration depth is the best choice for X-band electron spin resonance applications. They also detect partial release of the Meissner current at the vortex penetration field, suggesting that the interaction between vortices and the Meissner current near the edges is essential to understand the magnetic field dependence of the resonator properties.

Most recent results: Very recently the absorption of microwaves by granular, fine powders of MgB2 and K3C60 type II superconductors was measured and found to be giant, much larger than in the normal state, for magnetic fields as small as a few % of the upper critical field, a quite unexpected result [55]. The authors had the care to experimentally have the grains separated in large distances, so no weak links could be formed and no microwave absorption at the Josephson junctions could take place since no JJ were present.

Yet the microwave absorption is very large, and it represent the case in which $\sigma_1 \gg \sigma_2$ [46]. We need to remember that σ_1 is due to the normal conductivity and σ_2 is the superconducting conductivity.

The effect is predicted by the theory of vortex motion in type II superconductors; however, its direct observation had been elusive due to skin-depth limitations; conventional microwave absorption studies employ larger samples where the microwave magnetic field exclusion significantly lowers the absorption. The authors show that the enhancement is observable in grains smaller than the penetration depth. A quantitative analysis on K3C60 in the framework of the Coffey-Clem (CC) theory (see the connection theory experiment in **Figure 11**) explains well the temperature dependence of the microwave absorption and also allows to determine the vortex pinning force constant P.

Dissipation and radiation of microwaves of the HTSCs have been measured since their discovery. Bednorz and Muller (the very same discoverers) and Blazey and many others [34–38] reported the modulated microwave absorption of YBaCuO and of several other HTSC families. The basic equipment used in these studies is pretty much like the one shown in **Figure 11**. The electron paramagnetic resonance equipments are well suited to carry on these Q and Δf determinations provided that a zero magnetic field unit that could cross the zero value of the magnetic field is incorporated to a EPR spectrometer, in addition to a low-temperature (helium and/or liquid nitrogen) system. The cuprates, with their $3d^9$ unpaired electron, were expected to elicit EPR signals, and in consequence, some microwave absorption by this means was also expected. And certainly, these signals and absorptions have been detected and reported many times [34–38]. But in addition, the cuprates show an important non-resonance absorption at zero magnetic field or near-zero magnetic field [34–37]. Now it is clear that the Cu unpaired, bounded electron is not responsible of such microwave absorption. Bednorz and Muller and other authors have identified as the main absorbers, in the cuprates, the fluxoids, their dissipative dynamics and the loss processes in the weak links.

A strong microwave absorption at low magnetic fields is observed for a variety of metallic cuprates below their critical superconducting transition temperatures (T); using conventional electron paramagnetic resonance (EPR), instrumentation has been recorded [35]. The low-field microwave signal was investigated in the following high-temperature superconductors, YBaCuO (T = 95 K), GdBaCu (T = 95 K), etcetera. It has been investigated the usefulness of the technique as a practical nonintrusive method for screening potentially superconducting samples, with particular emphasis on sensitivity, the effect of microwave power, and the inherent problems of studying signals at small magnetic fields [32, 35].

7. Conclusions

The fundamental property of superconductors to absorb microwave energy in spite of the expulsion of static magnetic fields (Meissner effect) has been explored for various SC, and its use in superconducting microwave technology has been indicated. The experimental measurements of the absorption of microwaves in the form of ESR-like measurements were described, and emphasis was put on the quality factor as one central quantity to measure, and we stressed its great theoretical understanding. Q(T) and Q(H) magnetically modulated (or not) from microwave absorption were described, and few examples, spanning about 70 years of this field of research, were given. The knowledge gained on microwave losses can inform the design of new superconducting devices, SQC, JJs, SC qubits, microwave detectors, and/or radiators, operating in heavy microwave environments and/or as microwave devices themselves.

Author details

Rafael Zamorano Ulloa
Physics Department, Superior School of Physics and Mathematics, National Polytechnic Institute, Mexico City, Mexico

*Address all correspondence to: davozam@yahoo.com

References

[1] Tinkham M. Introduction to Superconductivity. 2nd ed. USA: McGraw-Hill; 1996. ISBN 0-07-064878-6

[2] Ashcroft N, Mermin D. Solid State Physics. NY: Saunders College; 1976. 738 p. ISBN 0-03-083993-9

[3] Annett J. Superconductivity, Superfluids and Condensates. NY: Oxford; 2004. 58 p. ISBN 0-19-850756-9

[4] Onnes H. Further experiments with liquid helium. D. On the change of electric resistance of pure metals at very low temperatures, etc. V. The disappearance of the resistance of mercury. Comm. Phys. Lab. Univ. Leiden; Through Measurement to Knowledge. 124; pp. 264-266. DOI: 10.1007/978-94-009-2079-8

[5] Meissner W, Ochsenfeld R. Ein neuer Effekt bei Eintritt der Supraleitfähigkeit. Die Naturwissenschaften. 1933;**21**:787-788. DOI: 10.1007/BF01504252

[6] London F, London H. The electromagnetic equations of the superconductor. Proceedings of the Royal Society A. 1935;**149**:71-78. DOI: 10.1098/rspa.1935.0048

[7] Pippard A, Bragg W. The surface impedance of superconductors and normal metals at high frequencies I. resistance of superconducting tin and mercury at 1200 Mcyc./sec. Proceedings of the Royal Society A. 1947;**191**:370-384. DOI: 10.1098/rspa.1947.0121

[8] Pippard A. Trapped flux in superconductors. Philosophical Transactions of the Royal Society A. 1955;**248**:97-129. DOI: 10.1098/rsta.1955.0011

[9] Bardeen J, Cooper L, Schrieffer R. Theory of superconductivity. Physics Review. 1957;**108**:1175-1204. DOI: 10.1103/PhysRev.108.1175

[10] Ginzburg V, Landau L. Collected Papers of L. D. Landau: On the Theory Superconductivity. Zh. Eksp. Teor. Fiz. 20, 1064. Oxford: Pergamon Press; 1965. 546 p. DOI: 10.1016/B978-0-08-010586-4.50078-X

[11] Abrikosov A, Zh. Eksp. Teor. Fiz. 32, 1442 (1957) (English translation: Sov. Phys. [JETP 5 1174 (1957)].) Abrikosov's original paper on vortex structure of type-II superconductors derived as a solution of G-L equations for $\kappa > 1/\sqrt{2}$; L.P. Gor'kov, Sov. Phys. JETP 36, 1364 (1959)

[12] AA Abrikosov's 2003 Nobel lecture: pdf file; V.L. Ginzburg's 2003 Nobel Lecture: pdf file

[13] Angulo-Brown F, Yépez E, Zamorano-Ulloa R. Finite-time thermodynamics approach to the superconducting transition. Physica A. 1993;**140**:431-436. DOI: 10.1016/0375-9601(93)90601-U

[14] London F. Superfluids. In: Macroscopic Theory of Superconductivity. Vol. I. New York: Dover Publications; 1961. 152 p

[15] Josephson B. Possible new effects in superconductive tunneling. Physics Letters. 1962;**1**:251-253. DOI: 10.1016/0031-9163(62)91369-0

[16] Landau L, Lifshitz E. The Classical Theory of Fields. 4th ed. Vol. 2. Butterworth-Heinemann: USSR; 1967. ISBN 978-0-7506-2768-9

[17] Kittel C. Introduction to Solid State Physics. 8th ed. USA: John Wiley & Sons, Inc.; 2005. ISBN 0-471-41526-X

[18] Superconducting Magnetic Energy Storage [internet]. Available

from: https://en.wikipedia.org/wiki/Superconducting_magnetic_energy_storage. [Accessed: 01 January 2019]

[19] Xing L, Jihong W, Dooner M, Clarke J. Overview of current development in electrical energy storage technologies and the application potential in power system operation. Applied Energy. 2015;**137**:511-536. ISSN 0306-2619

[20] Tixador P. Superconducting magnetic energy storage: Status and perspective. IEEE/CSC & ESAS Euro Superconductivity News Forum. 2008;(3). snf.ieeecsc.org

[21] SCMaglev [internet]. Available from: https://en.wikipedia.org/wiki/SCMaglev, [Accessed: 01 January 2019]

[22] Coates K. "High-Speed Rail in the United Kingdom" (PDF). Available from: https://web.archive.org/web/20110919101806/http:/namti.org/wp-content/uploads/2007/05/NAMTI-The-Lesson-From-TGV%E2%80%99s-HSR-Record-2007-5-1.pdf. [Retrieved: 01 January 2019]

[23] Klauda M, Kasser T, Mayer B, Neumann C, Schnell F, Aminov B, et al. Superconductors and cryogenics for future communication systems. IEEE Transactions on Microwave. 2000;**48**:1227-1234

[24] Military and Aerospace Electronics [Internet] McHal J. High-Temperature Superconductors Lighten Satellite Payload, September Issue, 1997. Available from: https://www.militaryaerospace.com/articles/print/volume-8/issue-9/news/high-temperature-superconductors-lighten-satellite-payloads.html. [Accessed: 01 January 2019]

[25] Superconducting Radio Waves Science and Technology [Internet]. Available from: https://www.helmholtz-berlin.de/media/media/grossgeraete/srf/00_what_ist_srf.pdf. [Accessed: 01 January 2019]

[26] Sucher M, Fox J. Handbook of Microwave Measurements. New York/London: Interscience Publishers: Polytechnic Press; 1963. 1145 p. https://www.amazon.com/Handbook-Microwave-Measurements-Max.../0470835389

[27] Lancaster M. Passive Microwave Device Applications of High-Temperature Superconductors. USA: Cambridge University Press; 1997. ISBN 0-521-48032-9(hp)

[28] Superconducting Radio Frequency [Internet]. Available from: https://en.wikipedia.org/wiki/Superconducting_radio_frequency. [Accessed: 21 March 2019]

[29] Weingarten W. Superconducting Cavities Basics. 1997. [Internet]. Available from: https://cds.cern.ch/record/308015/files/p167.pdf. CERN, Geneva, Switzerland. [Accessed: 21 March 2019]

[30] International Linear Collider [Internet]: Available from: https://en.wikipedia.org/wiki/International_Linear_Collider. [Accessed: 23 March 2019]

[31] Jackson D. Classical Electrodynamics. 3th ed. USA: John Wiley & Sons; 1999

[32] Zamorano R, Hernandez G, Villegas V. The interaction of microwaves with materials of different properties. In: Yeap KH, editor. Electromagnetic Fields and Waves. 1st ed. London: IntechOpen; 2019. ISBN 978-1-78923-956-0

[33] Superconductive Electronics Group. NIST. Fundamental Constants [internet]. Available from: https://www.nist.gov/pml/quantum-electromagnetics/superconductive-electronics [Accessed: 26 March 2019]

[34] Blazey KW, Muller KA, Bednorz JG, Berlinger W, Amoretti G, Buluggiu V, et al. Low-field microwave absorption

in the superconducting copper oxides. Physical Review B. 1987;**36**: 7241-7243

[35] Janes R, Ru-Shi L, Eduards P, Stevens AD, Symons MCR. Magnetic field dependent microwave superconducting cuprates. Journal of the Chemical Society, Faraday Transactions. 1998;**94**:3527-3536

[36] Pertile A, Lopez OA, Farach HA, Creswick RJ, Poole CP Jr. Model for low-field microwave absorption in granular type-II superconductors. Physical Review B. 1995;**52**:15475-15478

[37] Alvarez G, Zamorano R. Characteristics of the magnetosensitive non-resonant power absorption of microwaves by magnetic materials. Journal of Alloys and Compounds. 2004;**369**:231-234. DOI: 10.1016/j.jallcom.2003.09.058

[38] Wesche R. Physical Properties of High-Temperature Superconductors. 3th ed. UK: Wiley; 2015. ISBN 978-1-119-97881-7

[39] Clarke J, Wilhelm F. Superconducting quantum bits. Nature. 2008;**453**: 1031-1042. DOI: 10.1038/nature07128

[40] Schoelkopf R, Girvin M. Wiring up quantum systems. Nature. 2008;**451**:664-669. DOI: 10.1038/451664a

[41] MIT Technology News. The Record for High-Temperature Superconductivity has been Smashed Again. 2018, [Internet], Available from: https://www.technologyreview.com/s/612559/the-record-for-high-temperature-superconductivity-has-been-smashed-again/. [Accessed: 26 March 2019]

[42] Keimer B, Kivelson SA, Norman MR, Uchida S, Zaanen J. High Temperature Superconductivity in the Cuprates. ArXiv 2014, Available at: https://arxiv.org/ftp/arxiv/papers/1409/1409.4673.pdf. [Accessed: 26 March 2019]

[43] Ginsburg-Landau Theory [Internet] Available from: https://en.wikipedia.org/wiki/Ginzburg%E2%80%93Landau_theory. [Accessed: 24 March 2019]

[44] Superconductivity: The Meissner Effect, Persistent Currents and the Josephson effects. MIT Department of Physics. February 7, 2018. Available at: http://web.mit.edu/8.13/www/JLExperiments/JLExp39.pdf. [Accessed: 26 March 2019]

[45] Anderson W, Rowell M. Probable observation of the Josephson superconducting tunneling effect. Physical Review Letters. 1963;**10**:230-232. DOI: 10.1103/PhysRevLett.10.230

[46] Xiu G, Frisk-Kockumb Miranowicz A, Liua Y, Nori F. Microwave photonics with superconducting quantum circuits. Physics Reports. 2017:1:102-1:718. DOI: 10.1016/j.physrep.2017.10.002

[47] Shapiro S. Josephson currents in superconducting tunneling: The effect of microwaves and other observations. Physical Review Letters. 1963;**11**:80-82. DOI: 10.1103/PhysRevLett.11.80

[48] Bednorz JG, Müller KA. Possible high TC superconductivity in the Ba-La-Cu-O system. Zeitschrift für Physik B. 1986;**64**:189-193. DOI: 10.1007/BF01303701

[49] Wu MK, Ashburn JR, Torng CJ, et al. Superconductivity at 93 K in a new mixed-phase Y-Ba-Cu-0 compound system at ambient pressure. Physical Review Letters. 1987;**58**:908-910. DOI: 10.1103/PhysRevLett.58.908

[50] High-Temperature Superconductivity [Internet] Available at: https://en.wikipedia.org/wiki/High-temperature_superconductivity. [Accessed: 26 March 2019]

[51] Song C-L, Xue Q-K. Cuprate superconductors may be conventional after all. Physics. 2017;**10**:33-44. DOI: 10.1103/Physics.10.129

[52] Wertham R. Nonlinear self-coupling of Josephson radiation in superconducting tunnel junctions. Physics Review. 1966;**147**:255-263. DOI: 10.1103/PhysRev.147.255

[53] Eck E, Scalapino J, Taylor N. Physical review letters: Self-detection of the ac Josephson current. Physical Review Letters. 1964;**13**:15-19. DOI: 10.1103/PhysRevLett.13.15

[54] Tikhonov KS, Skvortsov MA, Klapwijk TM. Superconductivity in the presence of microwaves: Full phase diagram. Physical Review B. 2018;**97**:184516. DOI: 10.1103/PhysRevB.97.184516

[55] Csősz G, Márkus BG, Jánossy A, et al. Giant microwave absorption in fine powders of superconductors. Scientific Reports. 2018;**8**:11480-11485. DOI: 10.1038/s41598-018-29750-7

[56] Kwon S, Roudsari AF, et al. Magnetic field dependent microwave losses in superconducting niobium microstrip resonators. Journal of Applied Physics. 2018. p. 033903-124. DOI: 10.1063/1.5027003

Chapter 3

Bulk High-Temperature Superconductors: Simulation of Electromagnetic Properties

Ekaterina Kurbatova, Pavel Kurbatov and Mikhail Sysoev

Abstract

The chapter deals with the electromagnetic properties of bulk high-temperature superconductors (HTSs), which can be used in magnetic systems for various applications, in particular, in contactless magnetic suspensions. Magnetic levitation in HTS material has a different nature from permanent magnets. It is caused by induced superconducting currents inside the volume of material. Due to this, the levitation is self-stabilizing and does not require additional active control or mechanical stops in magnetic systems with HTS. HTS materials have nonlinear, anisotropic, hysteresis properties, which make the calculation of the superconducting devices very difficult. Here you can find a brief overview of existing approaches to modeling HTS materials by E-J characteristics. Authors propose the method of simulation intending for 3D numerical calculation, which represents the processes in HTS using two types of magnetic field sources – current and magnetization. The chapter focuses on the analysis of sources inside the superconducting material and their influence on an external magnetic field and levitation properties of HTS. In addition to simulations, the experimental studies of the force interactions between HTS bulks and permanent magnet are presented and compared with the calculations to verify the proposed mathematical models.

Keywords: high-temperature superconducting, model of HTS, bulk superconductor, the Meissner effect, simulation, numerical calculation, magnetic field, magnetic levitation

1. Introduction

Using bulk high-temperature superconductors (HTSs) is caused by the peculiarity of their electromagnetic properties. Due to the low resistivity, which is almost zero in the superconducting state, induced currents in HTS practically do not fade and reach significantly higher values than the conventional conductive materials.

According to Faraday's law, the induced superconducting currents prevent the change in the magnetic field; therefore, after the transition into the superconducting state, HTS traps and saves the magnetic field in which they were cooled. Based on this feature, it is possible to create full magnetic levitation characterized by self-stabilization in the directions of several degrees of freedom simultaneously. Unlike the passive magnetic bearing, where the interaction between permanent magnets provides the force, HTS suspensions do not require the additional active control or mechanical stops.

Most often, the schemes with the interaction of permanent magnets and bulk superconductors are used. In such constructions, two modes of HTS cooling are possible – zero-field cooling (ZFC) without external magnetic field and field cooling (FC) under the influence of the external field. In FC mode, HTS is cooled inside the magnetic system in the position that must be maintained. In this case, the HTS suspension is stable for any displacements. ZFC shows larger values of forces. However, it may lead to the instability of the system.

Creating of superconducting bearings [1–8], transportation systems (MagLev) [9–14], and other devices with HTS bulks requires the use of numerical calculation methods and specialized mathematical model describing their electromagnetic properties. Simulation of HTS bulks is a complex problem due to the necessity to consider the influence of nonlinear properties of superconducting material at work in magnetic fields and their interaction with the magnetic system.

In general, the following phenomena should be considered when modeling bulk HTS:

- transferring between the superconducting and the normal states;
- nonlinear resistance;
- the Meissner effect (exclusion of the field during cooling);
- trapping of a magnetic field;
- anisotropy; and
- hysteresis.

In this chapter, we briefly overview the existing models of HTS and present the method for simulation, which allows taking into account the various aspects of the bulk HTS properties listed above.

The possibilities of the proposed models are shown on simulation examples and compared with the experimental measurements. An example of a simulation of the magnetization process illustrates the distribution of magnetic field sources inside a superconductor. The behavior of HTS at different cooling modes, including the trapping of the magnetic field and the Meissner effect, is explained based on the analysis of simulation results.

The verification of the proposed method of simulation is performed by the comparison with measurements. Here we present the experimental study of the levitation force between a permanent magnet and HTS bulks, which is the most important in terms of the use of HTS bulks in magnetic suspensions and other levitation systems.

2. E-J characteristics of HTS

As a type of conductive material with a nonlinear resistivity, HTS is usually simulated by the E-J characteristics. The simplest model for the calculation of superconductors is *Critical State Model* [15–18]: when

$$\begin{cases} \mathbf{J} = J_C \dfrac{\mathbf{E}}{|\mathbf{E}|}, & when \ \ |\mathbf{E}| \neq 0 \\ \dfrac{\partial \mathbf{J}}{\partial t} = 0, & when \ \ |\mathbf{E}| = 0 \end{cases} \tag{1}$$

where **J** is the current density vector, **E** is the vector of the electric field strength, and J_C is the critical value of current density.

In this model, HTS is characterized by the E-J dependence at which the current in a superconductor cannot exceed a certain critical value J_C, and as the value of J_C is not achieved, the electric field is equal to zero (**Figure 1(a)**).

Determination of the critical current is an important problem in the simulation of HTS. The most common approaches are the Bean [15] and Kim [16] models. The Bean model assumes that $J_C = const$, and it is determined only by the properties of the superconductor. The studies in this area showed that J_C depends on external factors, in particular on the magnetic field. Thereby, the Kim model describes this influence as follows:

$$J_C = J_{C0} \frac{B_0}{B_0 + B}, \tag{2}$$

where B is the magnetic flux density, and J_{C0} and B_0 are the constants determined by the properties of a material. In general, J_{C0} and B_0 also depend on the temperature.

Extended Critical State Model [19, 20] was proposed to take into account the appearance of resistivity in case of exceeding J_C, caused by flux motion. This process can be modeled as a transition of a superconductor in the normal state (**Figure 1(b)**). For this purpose, the behavior of the superconductor in the flux-flow region [21, 22], when $J > J_C$, is defined using the additional equation:

$$E = \rho_f (J - J_C), \tag{3}$$

where ρ_f is the effective electrical resistivity. Thus, in this model, there are two areas, or conditions, in which the superconductor can be:

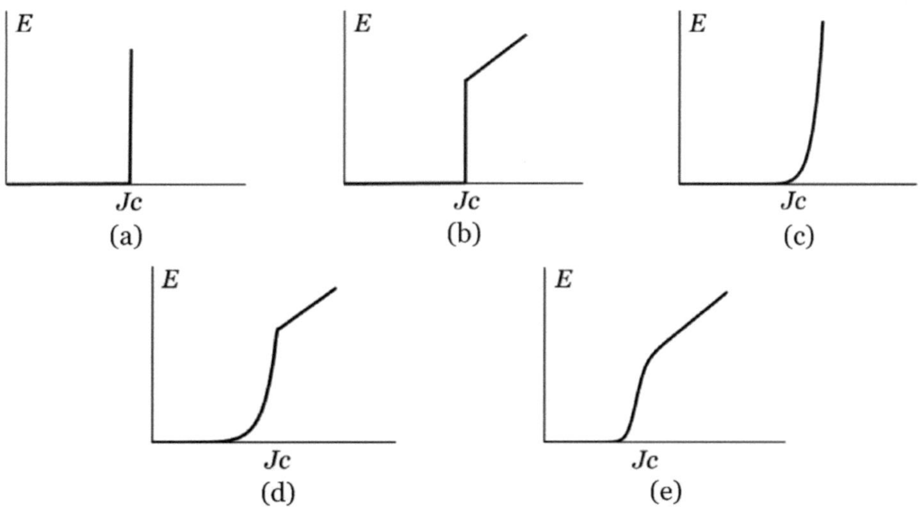

Figure 1.
E-J characteristics of HTS: (a) critical state model; (b) extended critical state model; (c) power law; (d) flux-flow and flux-creep model; and (e) hyperbolic model.

$|J - J_C| < J_C$ – superconducting state corresponding to the critical state model and
$J > J_C$ – appearance of the resistance caused by flux motion.

Critical current density and the effective electrical resistivity are the functions of the magnetic field strength:

$$J_C = J_0 (H_j/H)^n, \tag{4}$$

$$\rho_f = \rho_0 (H/H_{C2})^m, \tag{5}$$

where J_0 and H_j are the constants determined by the properties of material, ρ_0 is the resistivity of the superconductor in the normal state, and H_{C2} is the value of the second critical field.

The critical state model and its extension described above use linear approximation. This is an idealized representation of HTS properties. The real E-J dependence has nonlinear form, due to inhomogeneity of material and physical processes associated with the creep of fluxes, occurring with the increase of current. Such kind of dependence can be described using the *Power Law Model* [23, 24], which is a smooth approximation of the critical state model as shown in **Figure 1(c)**:

$$E = E_C \left(\frac{J}{J_C} \right)^n, \tag{6}$$

where E_C is the critical electric field strength, and n is the power law exponent that mainly influences the simulating properties [25].

Similar to J_C (Eq. (2)), n value depends on the magnetic field:

$$n = n_{C0} \frac{B_0}{B_0 + B}, \tag{7}$$

where n_{C0} and B_0 are the constants determined by the properties of material.

The power law model is the most commonly used for numerical calculations of superconducting materials. However, experimental determination of n value causes difficulties for HTS bulks [26].

One of the models that most accurately and completely describe the behavior of HTS is *Flux-Flow and Flux-Creep Model* [27, 28]. It divides the different types of conductivity in a high-temperature superconductor (**Figure 1(d)**):

$$E = \begin{cases} E_C \dfrac{\sinh\left(\beta J / J_C \right)}{\sinh(\beta)}, & when \quad J \le J_C \\ E_C + \rho_{ff}(J - J_C), & when \quad J > J_C \end{cases}, \tag{8}$$

where ρ_{ff} is the flux-flow resistivity, parameter β characterizes the pinning:

$$\beta = U_0 / k_B T, \tag{9}$$

where U_0 is the pinning potential, k_B is the Boltzmann constant, and T is the temperature.

When $J \le J_C$, a superconductor is in the flux-creep region [29]. When $J > J_C$, both flux-creep and flux-flow [21, 22] effects are present in material simultaneously.

The critical current density in Eq. (8) depends on the magnetic induction according to the Kim model mentioned above. And the flux-flow resistivity is defined from the Bardeen-Stephan model

$$\rho_{ff} = \rho_{f0}\frac{B}{B_\rho}, \tag{10}$$

where ρ_{f0} and B_ρ are the constants at a given temperature.

In some works, the electrical resistivity is represented as an approximating *Hyperbolic function* of magnetic field strength H, current density J, and temperature T [30, 31]:

$$\rho \;=\; 0.5\rho_0 \times \{1 + \text{th}\left[-(1 - T/T_C)(1 - H/H_C)(1 - J/J_C)/\delta_j\right]\}, \tag{11}$$

where ρ_0 is the electrical resistivity of the HTS material in the normal states (constant); J_C is the critical current density; H_C is the critical magnetic field strength; T_C is the critical temperature; and δ_j is the relative dispersion of the distribution of critical parameters in the inhomogeneous material.

E-J characteristic corresponding to Eq. (11) is shown in **Figure 1(e)**. This approximation is similar to the flux-flow and flux-creep model and allows to describe the properties of HTS in different regions.

For this model, the dependence of the critical current density on the magnetic field strength is determined by the power functions:

$$\begin{cases} J_C = \pm J_{C,\max} \cdot \left(1 - (H/H_c)^{\alpha}\right)^{\beta}, & \textit{when} \quad H \leq H_C \\ J_C = 0, \quad \textit{when} \quad H > H_C \end{cases}, \tag{12}$$

or

$$J_C = \pm J_{C,\max} / \left(1 + (H/H_c)^{\alpha}\right)^{\beta}, \tag{13}$$

where $J_{C,\max}$ is the maximum value of the critical current density at a given temperature, and α and β are the model parameters (real positive numbers). Variation of α and β allows us to change the form of the $Jc(H)$ dependence in a wide range. The maximum value of the critical current density is defined by the following temperature dependence:

$$J_{C,\max} = J_{C0}\left(1 - (T/T_C)^2\right)^n, \tag{14}$$

$$H_C = H_{C0}\left(1 - (T/T_C)^2\right)^m, \tag{15}$$

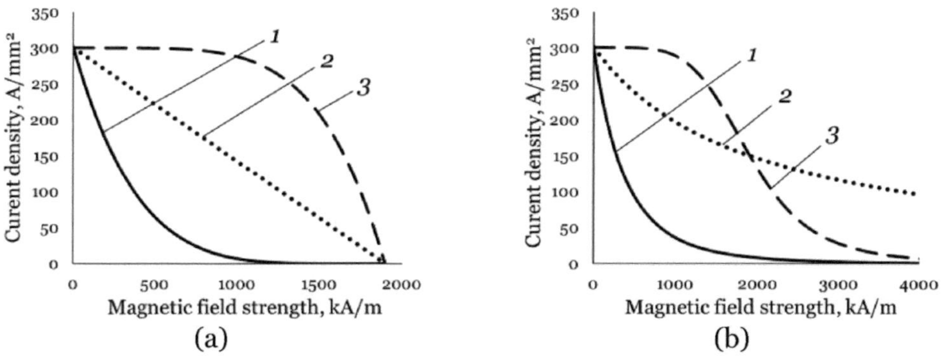

Figure 2.
Dependencies of the critical current density on the magnetic field strength at a constant temperature: (a) according to Eq. (12) and (b) according to Eq. (13). 1 – α = 1, β = 5; 2 – α = 1, β = 1; and 3 – α = 5, β = 1.

where J_{C0} and H_{C0} are the critical current density and critical magnetic field strength extrapolated to $T = 0$; Tc is the critical temperature, at which superconductivity is lost; and n and m are the positive real numbers.

Figure 2 illustrates the dependencies of the critical current density on the magnetic field strength at a constant temperature, according to Eqs. (12) and (13) with different coefficients α and β and the following critical parameters: $J_{C,\max} = 300$ kA/mm^2 and $H_C = 2000$ kA/m.

3. Method for simulation of HTS

The manufacturing technology of high-temperature superconducting materials determines their structure and electromagnetic properties. In general, mathematical models of the properties of such materials should describe not only the flow of conductivity currents, so-called "transport currents," but also the possible existence of small volume-distributed regions having magnetic properties similar to a nonlinear diamagnetic material. These regions can be represented as isolated superconducting regions in which circulating currents interact with the magnetic field according to a certain law as microcurrents in magnetic materials.

For numerical calculations, the distribution of superconducting currents can be approximated by the complex current vector models with two basic components: the transport current density $\mathbf{J}$ and the magnetization $\mathbf{M}$. To analyze an electromagnetic field, the sources $\mathbf{J}$ and $\mathbf{M}$ are represented in a volume of HTS by continuous functions of magnetic field parameters, that is, averaged at a macroscopic level.

The transport current $\mathbf{J}$ is induced to oppose the applied field in accordance with Faraday's law [32], which can be written for $\mathbf{J}$ using the relation $\mathbf{E} = \rho\mathbf{J}$

$$\mathbf{J} = -\frac{1}{\rho}\left(\frac{d\mathbf{A}}{dt} + \nabla\varphi\right), \tag{16}$$

where ρ is the electrical resistivity as the function of the magnetic field strength H, current density J, and temperature T. The electrical resistivity in superconductors is usually anisotropic and is represented by the resistivity tensor. Here we use the hyperbolic function Eq. (11), which describes the properties of HTS in a wide range and has functional customization. However, for the region without flux-flow resistance, the power law (Eq. (6)) will show similar results.

The vector magnetic potential $\mathbf{A}$ is the time-dependent function of magnetic field sources $\mathbf{J}$ and $\mathbf{M}$

$$A(t) = \frac{\mu_0}{4\pi}\left[\int\limits_V \frac{\mathbf{J}(t) + \nabla\times\mathbf{M}(t)}{r}dV - \oint\limits_S \frac{\mathbf{n}\times\mathbf{M}(t)}{r}dS\right] + \mathbf{A}^{\text{ext}}(t), \tag{17}$$

where $\mathbf{A}^{\text{ext}}$ is the vector potential created by the external sources.

The scalar electric potential φ determines the potential component of the electric field strength

$$\mathbf{E}^p = -\nabla\varphi = -\frac{1}{4\pi\varepsilon_0}\oint\limits_S \frac{\xi(t)\mathbf{r}}{r^3}dS + \mathbf{E}^{\mathrm{p,ext}}, \tag{18}$$

where ξ is the density of the electric charges induced on the surface of a superconductor, and $\mathbf{E}^{p,ext}$ is the electric field strength created by the external sources. The electric charge density is determined using the integral boundary equations.

Detailed information about Eqs. (17) and (18) can be found in Ref. [32].

The magnetization is determined from the stationary state equations and properties of material

$$\begin{cases} \mathbf{B} = \nabla \times \mathbf{A} \\ \mathbf{B} = \mu_0(\mathbf{H} + \mathbf{M}) \end{cases}. \tag{19}$$

Magnetization is associated with another type of source-related currents circulating in the isolated regions as mentioned above. These currents have the same properties as transport current in terms of depending on the temperature and magnetic field strength but have the form of small local vortex structures with the magnetic moment vectors **m.** For modeling the contribution of these local currents, they can be represented as the distribution of superconducting solenoids, which length is much more than its diameter.

If we consider such long solenoid, the intrinsic magnetic field strength inside it is close to uniform and equal to the linear current density. Thus, it is assumed that the local current structures are equivalent to uniformly magnetized elements along the axis. The own demagnetizing magnetic field strength of the solenoid is zero.

As a parameter of the material properties, the magnetization M is the density of magnetic moments of the superconducting solenoids [33]. Until the magnetization M in a superconductor is less than the value of the critical magnetization Mc, the magnetic flux density inside the superconducting solenoids remains constant and equal to the previously acquired value. It means that in the region $M < Mc$, the magnetization is changed according to the equation for a diamagnetic material

$$\frac{dM}{dH} = -1. \tag{20}$$

Due to the relation with the superconducting currents, the critical magnetization is characterized by the nonlinear dependence on the magnetic field like in Eqs. (12) and (13)

$$\begin{cases} M_C = \pm M_{C,\max} \cdot (1 - (H/H_{CM})^{\alpha})^{\beta}, \quad when \quad H \leq H_{CM} \\ M_C = 0, \quad when \quad H > H_{CM} \end{cases}, \tag{21}$$

or

$$M_C = \pm M_{C,\max} / (1 + (H/H_{CM})^{\alpha})^{\beta}, \tag{22}$$

where $M_{C,\max}$ is the maximum value of critical magnetization at a given temperature, and H_{CM} is the critical magnetic field strength for magnetization. The temperature dependencies of the critical magnetization and critical magnetic field strength are the same as for the hyperbolic model in Eqs. (13) and (14).

The model parameters α and β for the magnetization may differ from their values in the equations for the current density. **Figure 3** presents the dependencies of the maximum critical magnetization at a constant temperature according to Eqs. (21) and (22) with different coefficients α and β and the following critical parameters: $M_{C,\max}$ = 500 kA/m and H_{CM} = 2000 kA/m.

Figure 4 shows the magnetization curve of a superconductor obtained using Eqs. (20) and (21). If HTS has been transferred into the superconducting state without an external magnetic field (ZFC), then the magnetization in the initial state is zero (point 0). With increasing magnetic field strength, the magnetization increases in the opposite direction to satisfy the condition of an ideal diamagnetic $M = -H$ (magnetic induction inside the diamagnetic equals zero) until point 1. From this point, the magnetization is limited by the critical curve $M_C = f(H)$ up to $M_C = 0$ at point 6. Changing the magnetic field in the range $-M_C < H < M_C$ does not cause the trapping of a magnetic field, and magnetization is defined by the linear region of the characteristic. If the magnetic field strength decreases from the point on the critical line (point 2), then the magnetization will be defined from the condition of the constant value of magnetic induction B_2 as $M = B_2/\mu 0\text{-}H$, where B_2 is the value of magnetic flux density at point 2. At point 3, the magnetization reaches the critical value on the upper part of the magnetization curve, and furthermore, it will be determined by the critical curve $M_C = f(H)$ up to point 7. If the magnetic field decreases at point 4, the new value of magnetic induction B_4 will remain constant until the critical magnetization at point 5. At field cooling (FC), the magnetization changes similarly, but the initial point is determined by the magnetic field $H \neq 0$ either on the initial line 0–1 or on the critical curve 1–6.

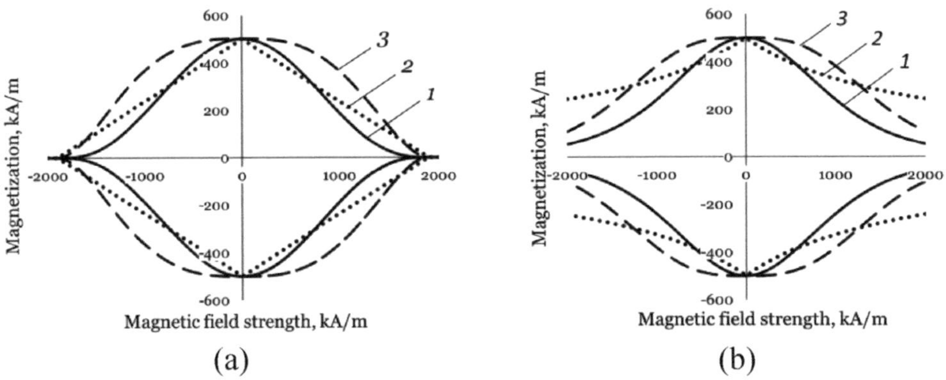

Figure 3.
Dependencies of the maximum critical current density on the magnetic field strength at a constant temperature: (a) according to Eq. (21) and (b) according to Eq. (22). $1 - \alpha = 2, \beta = 3$; $2 - \alpha = 1, \beta = 1$; *and* $3 - \alpha = 3, \beta = 2$.

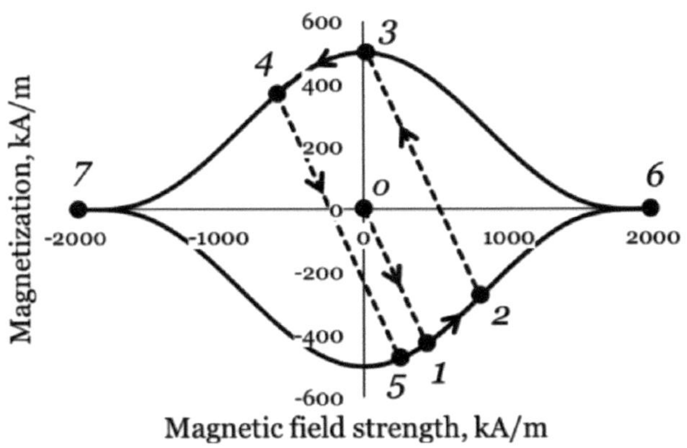

Figure 4.
Changing the magnetization under variation of a magnetic field (magnetization curve of HTS).

We assume that in a certain small volume of the superconductor, there are a sufficiently large number of solenoids, which may have different parameters of the critical magnetization dependencies. Inhomogeneity of the magnetization properties is taken into account using the probabilistic distribution laws for the mentioned parameters. **Figure 5** presents the comparison of the volume-averaged magnetization determined for a superconducting material with homogeneity and inhomogeneity properties. In the first case, all superconducting solenoids are identical. In the second case, solenoids have different critical parameters and, accordingly, create different magnetic moments.

The spatial orientation of the solenoids defines the anisotropic properties of the magnetization. $M(H)$ characteristics in **Figure 5** represent the properties along the axis of the solenoids, which have identical directions in a volume of superconductor. The orthogonal components of the magnetic field strength do not create magnetization. This corresponds to the usually used assumption that currents flow in the plane perpendicular to the anisotropy axis.

In general case, the anisotropic properties can be simulated using the probability distribution of both the values of critical parameters and directions. **Figure 6** shows an example of the dependence of the magnetization along the main axis of anisotropy α as a function of the components of the magnetic field strength along the axes α and β.

The idea of combining the models for current and for magnetization in the simulation of HTS bulks is as follows. Applying the external magnetic field after

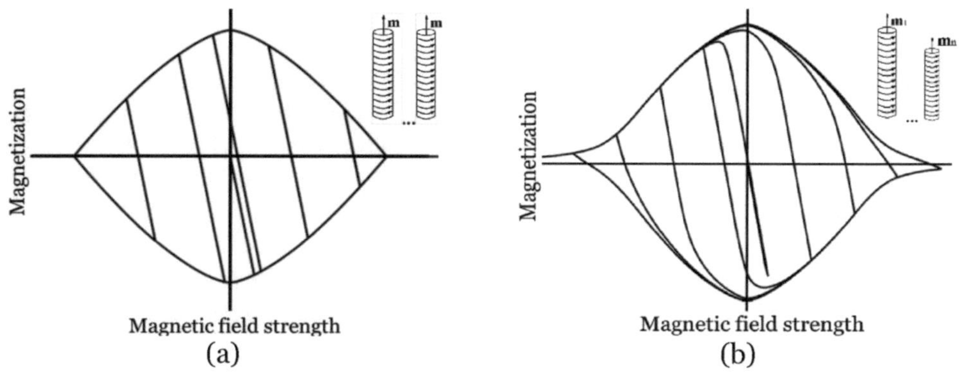

Figure 5.
Magnetization curve of HTS bulk with homogeneity (a) and inhomogeneity (b) properties.

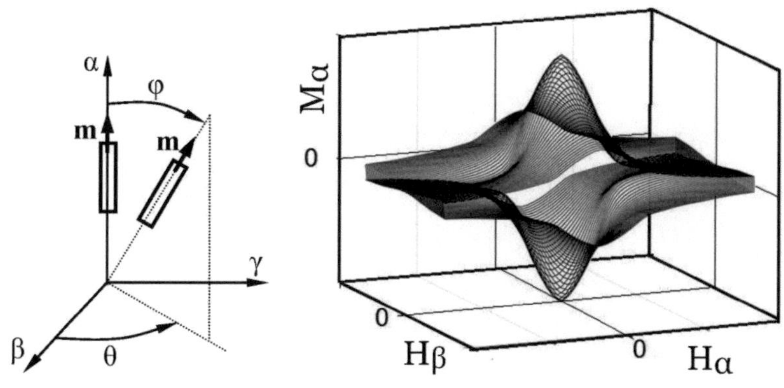

Figure 6.
Magnetization model with anisotropy.

cooling the superconductor at ZFC mode leads to the appearance of the induced transport currents. These currents are defined by both the external magnetic field and the own magnetic field of superconductor created by the resulting magnetization and induced currents. During the cooling process at FC mode, the magnetization occurs to expel the magnetic field from the volume of HTS as in type I superconductors. Changes of the resulting magnetic field due to the magnetization causes induced currents to flow according to Faraday's law. In this case, currents and magnetization create magnetic fields that try to compensate for the influence of each other. If the compensation of magnetic fields does not full, there will be the partial exclusion of the magnetic field from the superconductor. Thus, it is possible to simulate the partial Meissner effect at FC mode.

4. Simulation of J and M sources in HTS bulk

This section presents the analysis of the distribution of field sources in a bulk high-temperature superconductor. Computational model for analyzing the interaction of HTS bulk with the external magnetic field, which is shown in **Figure 7**, illustrates the simulation of superconductor using the proposed mathematical models for two types of electromagnetic field sources J and M. In this example, superconducting disk 1 is placed in the magnetic field created by the DC coil 2. Changing the current in the coil provides two different types of transferring HTS disk into the superconducting state: zero-field cooling (ZFC) in the absence of a magnetic field and field cooling (FC) in the presence of a magnetic field.

MMF in the coil is set as a time function (**Figure 7(c)**) depending on the cooling mode. In ZFC, the calculation starts with zero MMF, after which the current linearly increases until the maximum value and then goes back to zero. In FC, the initial MMF is equal to the maximum value. It remains constant for 1 s and then decreases to zero. This delay does not affect the simulation and had been used only for the convenience of comparison between calculated results at ZFC and FC modes.

The process of magnetizing the HTS disk allows us to analyze the behavior of different sources, including the features of distribution inside the volume of superconductor and their influence on the magnetic field near the disk. Here, we carried out the calculations with three different models of HTS properties: model only for current, model only for magnetization, and combined model with current and magnetization. Despite the fact that both magnetization and current describe the motion of charges inside an HTS, that is, superconducting currents, as mathematical models of magnetic field sources, they have different properties and obey

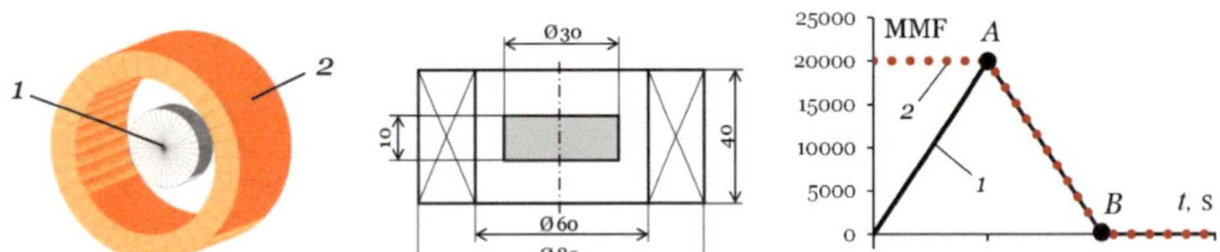

Figure 7.
Computational model for analysis of the interaction HTS bulk with the external magnetic field: (a) 3D model for calculation: 1 – HTS disk, 2 – field coil; (b) dimensions of the elements; and (c) MMF in the coil: 1 – ZFC, 2 – FC.

different laws. This causes relevant differences in the magnetizing process, as shown below.

Parameters of the mathematical model for M and J used for simulation in this section are given in **Table 1**. Temperature is assumed to be a constant value of 77 K, and the critical temperature is 93 K.

Figure 8 presents the distribution of magnetic field sources in a section of HTS disk in the case of ZFC. Results are presented for two specific time points: point A – at the maximum value of MMF (t = 1.0 s) and point B – at zero MMF (t = 2.0 s). The current density was calculated using the model only for the current density (Eqs. (11) and (12)); the magnetization was obtained from the calculation with the model only for the magnetization (Eqs. (20) and (21)).

Parameter	Units	Value
Critical current density, $J_{C,max}$	A/mm^2	300
Critical magnetic field for current, H_C	kA/m	2300
α for critical current	—	1
β for critical current	—	2
Dispersion δ	—	0.01
Critical magnetization, $M_{C,max}$	kA/m	1000
Critical magnetic field for magnetization, H_{CM}	kA/m	2300
α for magnetization	—	1
β for magnetization	—	2

Table 1.
*Parameters of the mathematical models for J and M used in simulation in **Figure 7**.*

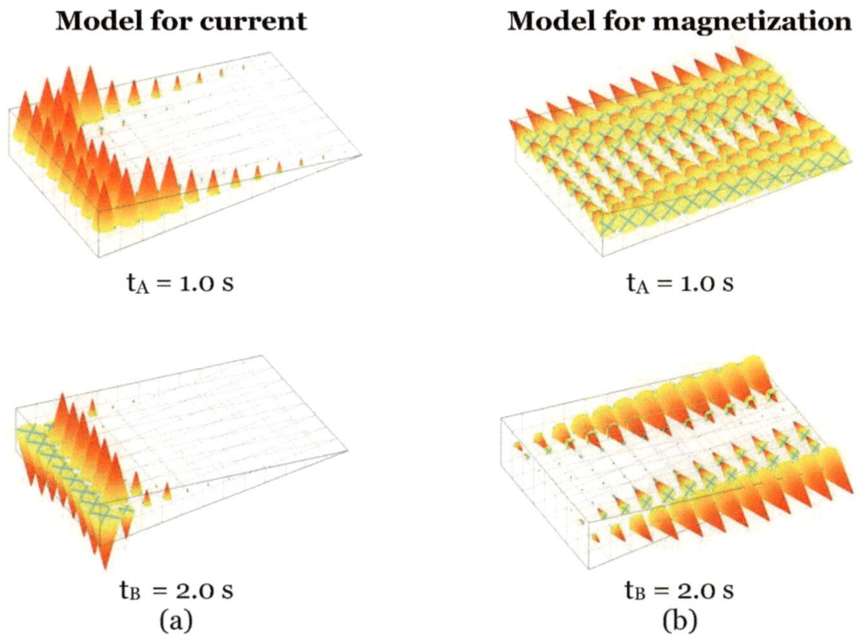

Figure 8.
Distribution of the magnetic field sources in a section of the HTS disk at ZFC: (a) model for current and (b) model for magnetization.

After transferring into the superconducting state, HTS prevents the change of the magnetic field inside its volume. At point A, current and magnetization try to compensate the magnetic field of the coil. However, due to the nonlinear properties of HTS, exceeding a certain value of the external field causes the magnetic field to penetrate inside the superconductor. Therefore, even after ending the current impulse at point B, both magnetic field sources save the trapped magnetic field. The maximum value of the external magnetic field, which can be compensated by HTS bulk, depends on the critical parameters of the mathematical models (J_C and M_C).

Figure 9 shows the distribution of magnetic field sources inside HTS in the case of ZFC obtained from the calculation using the combined model at similar time points. In the combined model, both magnetic field sources exist simultaneously and affect each other.

The combined model reproduces the features of the magnetizing process after ZFC that is compensating and trapping the external magnetic field as well as conventional models for current. The impact of an additional field source in the form of magnetization leads to a change in the distribution of current density inside the volume of HTS compared with the results of the model only for current. Currents in **Figure 8** are induced on the side surface of the disk over the entire height and gradually penetrate the volume. In contrast, in the combined model, the magnetization vectors also take part in compensation of the magnetic field, and currents flow only on the end faces of the disk filling the volume of HTS from the outer radius. The induced currents also affect the distribution of magnetization, which differs from that shown in **Figure 8**. This is especially observed at point B when HTS traps a part of the external magnetic field.

Distribution of the magnetic field near HTS disk is defined by the type and localization of magnetic field sources. **Figure 10** presents the calculated distribution of the axial H_Z and radial H_ρ components of the magnetic field strength along the radius on a distance of 1 mm above the superconductor at ZFC. Dependences for three considered models of HTS properties are shown at two specific time points,

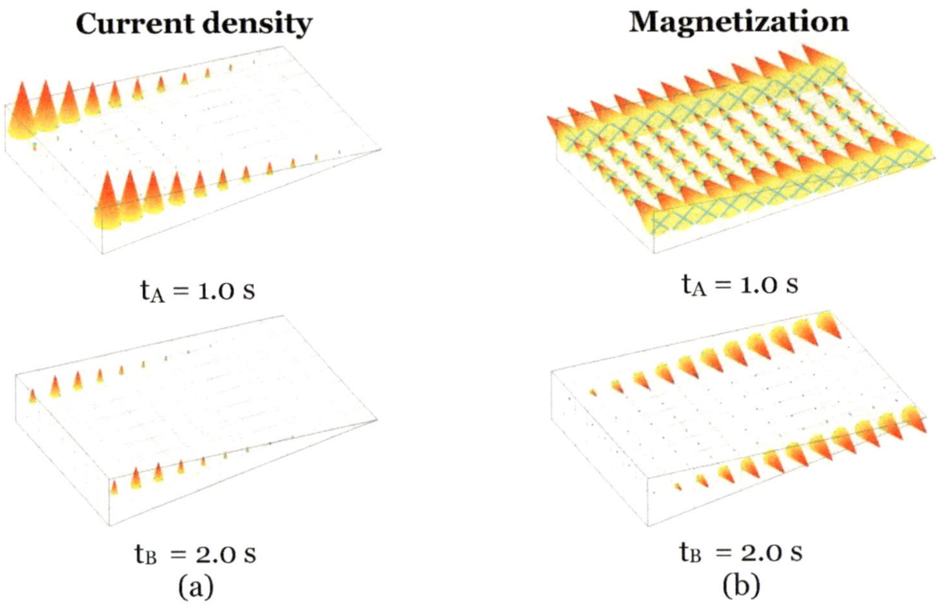

Figure 9.
Distribution of the magnetic field sources in a section of the HTS disk at ZFC using the combined model: (a) current density and (b) magnetization.

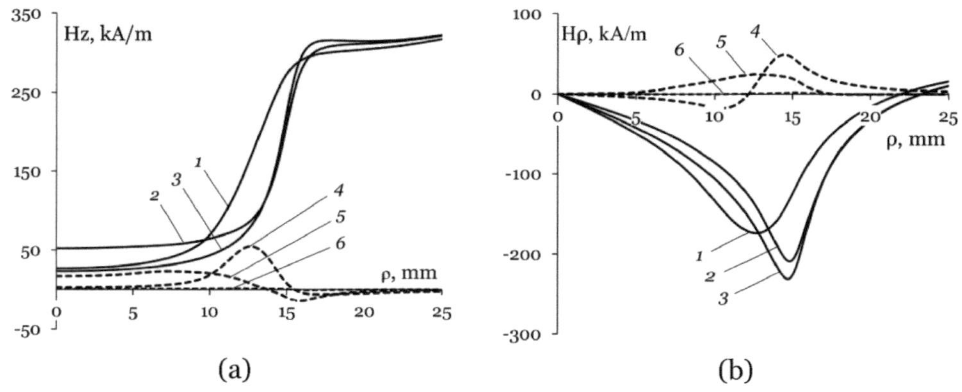

Figure 10.
Distribution of the axial (a) and radial (b) components of the magnetic field strength along the radius of HTS disk at ZFC: 1 – model for current at t_A = 1.0 s; 2 – model for magnetization at t_A = 1.0 s; 3 – combined model at t_A = 1.0 s; 4 – model for current at t_B = 2.0 s; 5 – model for magnetization at t_B = 2.0 s; and 6 – combined model at t_B = 2.0 s.

corresponding to above presented distributions of magnetic field sources: the maximum MMF (point A) – lines 1–3, and zero MMF (point B) – lines 4–6.

The obtained dependencies show that the model for magnetization gives a more uniform distribution of the axial component of the magnetic field strength than the model for current when the magnetic field penetrates HTS disk (point A). At the same time, the model for magnetization has a higher maximum and a sharper change of the radial component of the magnetic field strength at the border of HTS disk. At point B, the trapped magnetic field has a significant difference caused by the nature of the magnetic field created by currents and magnetization. A more pronounced difference is observed at point B because at this time, there is no external magnetic field, which primarily defines the distribution of the magnetic field at point A.

The distribution of the magnetic field strength for the combined model at point A is transitional between the results of calculations using the model only for current and the model only for magnetization. Thus, it becomes possible to customize the model of HTS properties by choosing model parameters and introducing weighting factors defining the contribution of each magnetic field source on the properties of superconducting material. The final condition of HTS is defined by the resulting influence of the current and magnetization and achieved when both sources come to a balance point, which is determined by the parameters of their models. In considered example with the applied model parameters, the magnetization almost completely compensates the impact of the induced current and its magnetic field. This leads to the absence of a trapped magnetic field at point B.

A similar simulation was carried out for cooling the HTS bulk in the magnetic field of DC coil. **Figure 11** presents the distribution of magnetic field sources in a section of a superconducting disk in the case of FC. Results are presented for two specific time points, similar to ZFC: point A – before changing the MMF (t = 1.0 s) and point B – at zero MMF (t = 2.0 s). The current density was calculated using the model only for the current density (Eqs. (11) and (12)); the magnetization was obtained from the calculation with the model only for the magnetization (Eqs. (20) and (21)).

When type I superconductor is cooled in an external magnetic field (FC), the magnetic field is excluded from its volume. This phenomenon is called the Meissner effect. In type II superconductors, this effect appears partially; therefore, usually it is not taken into account in calculations. As you can see in **Figure 11**, the current is

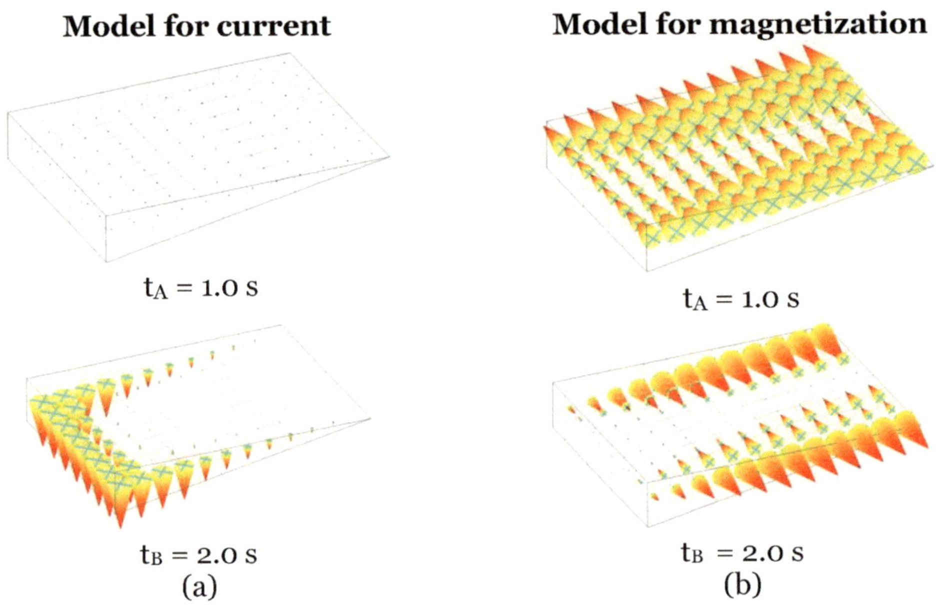

Figure 11.
Distribution of the magnetic field sources in a section of the HTS disk at FC: (a) model for current and (b) model for magnetization.

not induced at point A, since there is no change in the magnetic field during FC. In contrast to the model for current, the magnetization appears at the initial time and partially compensates the magnetic field inside the disk. However, due to the limitation of magnetization, most of the magnetic field penetrates the superconductor and then is trapped by it. At point B, both magnetic field sources are distributed in such a way as to trap the maximum of the magnetic field, which depends on the critical parameters of the mathematical models (J_C and M_C).

Figure 12 shows the distribution of the current density and magnetization vectors in a section of the HTS disk in the case of FC obtained from the calculation using the combined model with current density and magnetization at similar time points. Upon transition into the superconducting state, the magnetization and current appear simultaneously (point A). Magnetization provides the expulsion of the magnetic field from the volume of a superconductor (the partial Meissner effect). This causes a change in the magnetic field and inducing of superconducting current, which partially or wholly compensates the change of the magnetic induction in HTS. Therefore, unlike the model only for currents, when using the combined model, induced currents are observed even during FC. As for the trapped magnetic field at point B, it is provided by both sources, the distribution of which is similar to the separate mathematical models taking into account their influence on each other. Obtained distribution of currents and magnetization at FC mode for the combined model differs from the distributions of these sources in the ZFC mode. This mainly applies to currents, which occur on the side surface of the superconducting disk rather than on the end faces.

Figure 13 presents the calculated distribution of the axial H_Z and radial $H\rho$ components of the magnetic field strength along the radius on a distance of 1 mm above the superconductor at FC mode. Dependences for three considered models of HTS properties are shown at two specific time points, corresponding to the presented above distributions of magnetic field sources: before changing the MMF (point A) – lines 1–3 and at zero MMF (point B) – lines 4–6.

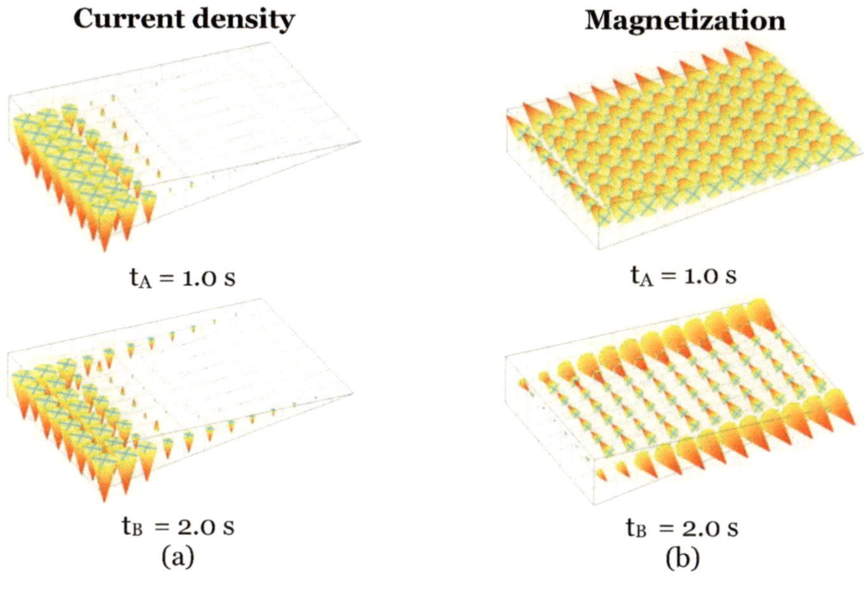

Figure 12.
Distribution of the magnetic field sources in a section of the HTS disk at FC using the combined model: (a) current density and (b) magnetization.

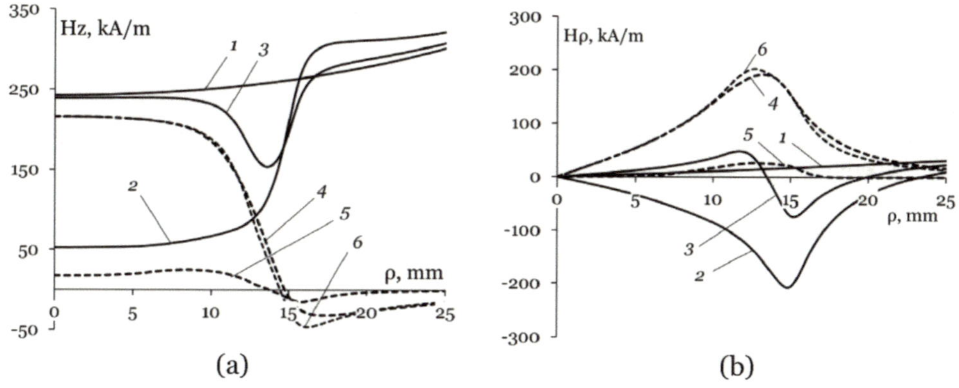

Figure 13.
Distribution of the axial (a) and radial (b) components of the magnetic field strength along the radius of HTS disk at FC: 1 – model for current at t_A = 1.0 s; 2 – model for magnetization at t_A = 1.0 s; 3 – combined model at t_A = 1.0 s; 4 – model for current at t_B = 2.0 s; 5 – model for magnetization at t_B = 2.0 s; and 6 – combined model at t_B = 2.0 s.

The obtained results of simulation indicate that the model for currents does not allow reproducing the Meissner effect in FC mode since the current in the HTS disk cannot be induced without changing the magnetic field. At point A, line 1 corresponds to the distribution of the magnetic field created by the current in the coil since the HTS disk does not affect it. In the model for magnetization, the resulting magnetic field due to the influence of magnetization significantly changes above the superconductor. Expulsion of the magnetic field from the volume of superconductor leads to a decrease in the axial component of the magnetic field strength and an increase in the radial component of the magnetic field strength compared with the coil field. The combined model shows the ability to control the nature of the source distribution and, accordingly, to adjust the Meissner effect and the redistribution of the magnetic field during FC mode quantitatively by choosing the model parameters.

At point B, the combined model and the model only for current show almost the same distribution of the trapped magnetic field. This is explained by the fact that the model for current mainly determines the trapped magnetic field in HTS at FC, while magnetization describes the changes of the magnetic field inside the superconducting material, which do not obey Faraday's law, including the Meissner effect.

5. Comparison with experimental measurements

Experimental study of the interaction force between a permanent magnet and HTS bulks, the so-called levitation force, was performed to provide data for the validation of the proposed mathematical models. Measurements were carried out on the laboratory equipment as shown in **Figure 14**.

Studied HTS sample 1 is placed and fixed inside the cuvette 2 using a nonmagnetic holder. For measuring the dependence of force on the gap between a permanent magnet and HTS, we use a moving rod 3, which is driven by a stepping motor. At the end of the rod, there is a fixed steel disk 4 holding the cylindrical permanent magnet 5 due to an attractive magnetic force. Weight 6 measures the force acting on the cuvette with HTS, and position sensor 7 shows the displacement of the permanent magnet during the experiment. Measuring equipment and a motor are connected to a computer that controls the experiment. For cooling, we used liquid nitrogen. Dimensions of the permanent magnet and two studied HTS samples (disk and ring) are shown in **Table 2**.

Force has been measured for two cooling modes: zero-field cooling and field cooling. During ZFC, the superconductor is cooled at the maximum gap of 50 mm. Then, the gap is installed at the initial measuring value of 29 mm. At FC, cooling process takes place at the gap of 3 mm. After cooling, the permanent magnet slowly moves vertically to or from HTS sample depending on the cooling mode.

Due to the properties of the high-temperature superconductors, the initial force curve differs from the subsequent ones. Therefore, the measurement along the

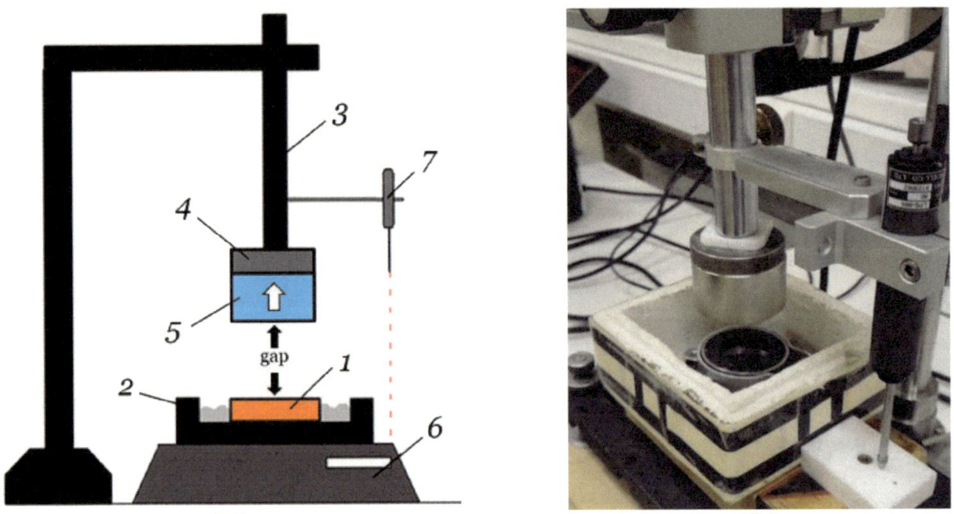

(a) (b)

Figure 14.
Experimental measurements of the levitation force: (a) scheme of the experiment and (b) – photo of the laboratory equipment. 1 – HTS sample, 2 – cuvette, 3 – moving rod, 4 – steel disk, 5 – permanent magnet, 6 – weights, and 7 – position sensor.

	HTS disk	HTS ring	Permanent magnet
External diameter, mm	46	46	50
Inner diameter, mm	—	30	—
Height, mm	7	7	30

Table 2.
Dimensions of the permanent magnet and HTS samples.

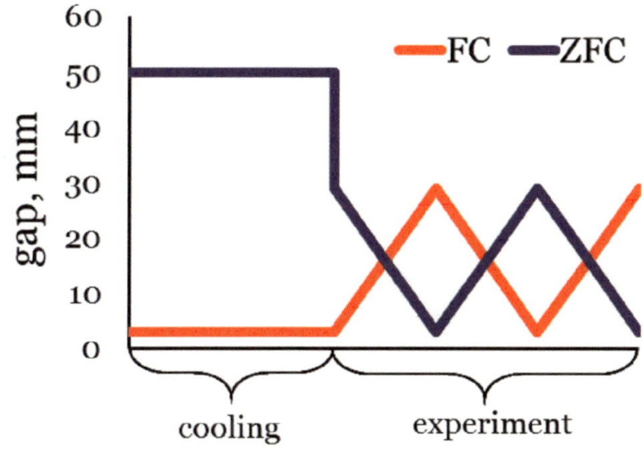

Figure 15.
Changing the gap during the experiment for FC and ZFC.

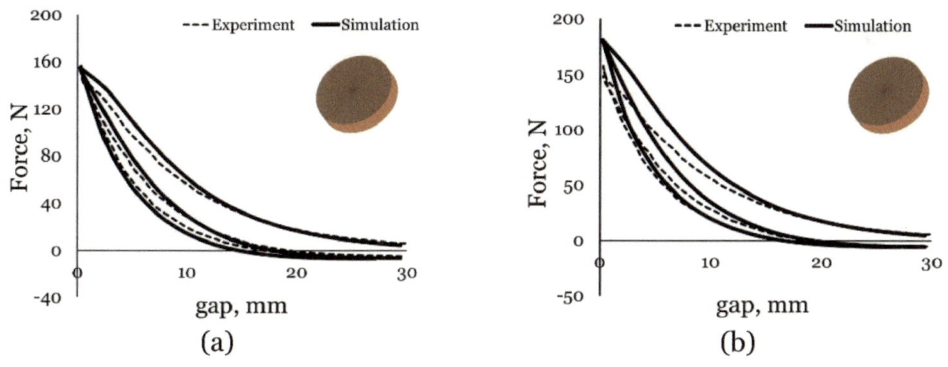

Figure 16.
Force of HTS disk for ZFC: (a) model for current and (b) combined model.

initial trajectory was performed twice to obtain stable characteristics. Thus, the change of the gap looks like this: 29–0.2–29–0.2 mm after ZFC and 3–29–3–29 mm after FC, as shown in **Figure 15**. With further cyclic loads, the obtained force dependencies will be repeated.

Experimental dependencies of the levitation force on the gap between the permanent magnet and HTS samples are compared with simulated results in **Figures 16–19**. The presented calculated characteristics have been obtained after selecting and optimizing the parameters of mathematical models described in Section 2 of this chapter.

Despite the fact that the approximation models have many parameters and due to this, they allow achieving excellent agreement with the experimental results in particular cases, defining the parameters is not an easy task on practice.

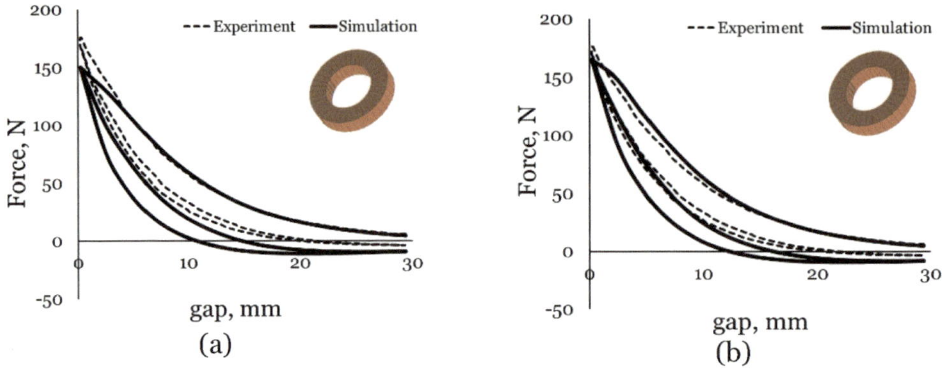

Figure 17.
Force of HTS ring for ZFC: (a) model for current and (b) combined model.

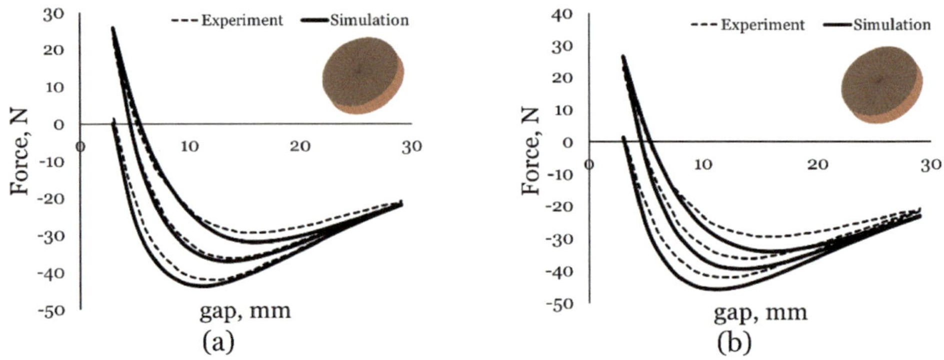

Figure 18.
Force of HTS disk for FC: (a) model for current and (b) combined model.

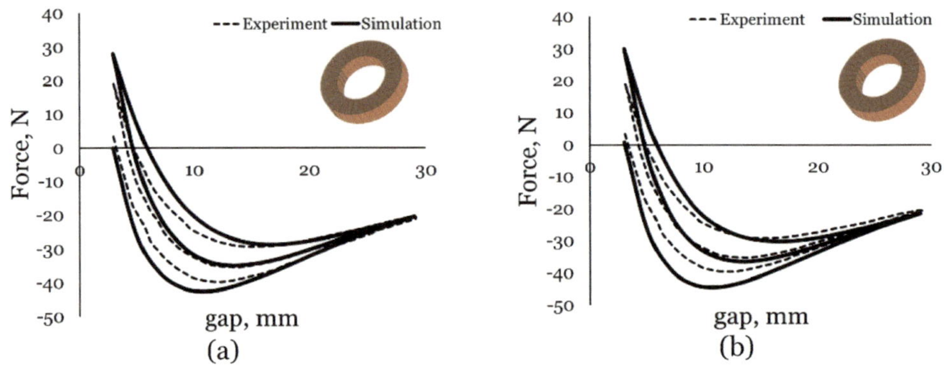

Figure 19.
Force of HTS ring for FC: (a) model for current and (b) combined model.

The difficulty lies in need to choose parameters that show a good approximation for both cooling modes – ZFC and FC.

The parameters selected based on the comparison with the measured results are presented in **Table 3**. Due to experimental confirmation, they can be considered as parameters of the studied HTS samples, and when studying several samples made of

Parameter	Units	Value
Critical current density, $J_{C,\max}$	A/mm^2	150
Critical magnetic field for current, H_C	kA/m	1100
α for critical current	—	1
β for critical current	—	2
Dispersion δ	—	0.003
Critical magnetization, $M_{C,\max}$	kA/m	80
Critical magnetic field for magnetization, H_{CM}	kA/m	1100
α for magnetization	—	1
β for magnetization	—	2

Table 3.
Parameters of the mathematical models for J and M from comparison with the experiment.

one material, as a characteristic of this material. Such a characteristic can be used in the design of devices based on bulk HTS with arbitrary sizes and shapes. To get closer to defining the parameters for material, rather than parameters for sample, two samples of different shapes were studied. Obtained parameters are suitable for both samples in both cooling modes.

During calculations and choosing of the model parameters, we assumed that both HTS samples have identical properties. However, the analysis of the experimental results of the experiment indirectly suggests some differences. The discrepancy between the calculated and experimental data was also influenced by inaccuracies of the experiment, in particular, some changes in the magnetization of the permanent magnet due to cooling by nitrogen at a minimum gap.

For illustrating the magnetization impact, all experimental dependencies were calculated for two models: model for current and combined model. At ZFC, magnetization causes the increase of the force at the minimal gap. At FC, magnetization shows the initial force at the cooling point, which is provided by the Meissner effect. In general, both models show good approximation with the experiment and can be recommended for analysis of magnetic systems with HTS bulk. As for magnetization, it should be applied in the system with a low gap, where it signally affects the characteristics of the system (force or trapped magnetic field), especially for FC.

Defined parameters, and the method for their determination, have limitations. Since they were selected and confirmed using experimental results at certain magnetic field range, their use is reliable only within the same field values.

6. Conclusion

The chapter discusses the methodology for modeling the electromagnetic properties of bulk HTS designed for the numerical calculation of magnetic systems. The proposed method is based on applying the two types of sources – current field and magnetization. The current properties are determined by nonlinear resistance. Magnetization is a model of local currents that can occur in a material, for example, upon transition into superconducting state. These currents do not obey the law of electromagnetic induction (in modeling). Their distribution in the volume can be represented as a characteristic of nonlinear magnetization.

Using the proposed models, we simulated the interaction of HTS bulk with external magnetic field after zero-field and field cooling. Distribution of the magnetic field sources inside a superconductor and magnetic field strength near it were obtained and analyzed. As shown in simulation results, the combination of two sources allows to widely regulate the properties of HTS and expand the possibilities of conventional methods, including the modeling of the Meissner effect.

For verification of the proposed method, we compare the results of simulation with the experimental measurements of the levitation force. Comparison shows the good agreement between calculated and measured results, which confirms the possibility to expand and clarify the approximation models describing the electromagnetic properties of bulk high-temperature superconductor.

Author details

Ekaterina Kurbatova[1]*, Pavel Kurbatov[1] and Mikhail Sysoev[2]

1 National Research University "Moscow Power Engineering Institute", Moscow, Russia

2 Bauman Moscow State Technical University (National Research University), Moscow, Russia

*Address all correspondence to: kurbatovaep@mail.ru

References

[1] Werfel FN, Floegel-Delor U, Rothfeld R, Riedel T, Goebel B, Wippich D, et al. Superconductor bearings, flywheels and transportation. Superconductor Science and Technology. 2012;**25**:014007. DOI: 10.1088/0953-2048/25/1/014007

[2] Krabbes G, Fuchs G, Canders WR, May H, Palka R. High Temperature Superconductor Bulk Materials: Fundamentals - Processing - Properties Control - Application Aspects. Wiley: Weinheim; 2006. p. 311. DOI: 10.1002/3527608044

[3] Coombs T. Bulk high temperature superconductor (HTS) materials. In: Melhem Z, editor. High Temperature Superconductors (HTS) for Energy Applications. Cambridge: Woodhead Publishing; 2012. pp. 101-139

[4] Kurbatova E. Comparative analysis of the specific characteristics of the magnetic bearings with HTS elements. IEEE Transactions on Applied Superconductivity. 2018;**28**:5207704. DOI: 10.1109/TASC.2018.2824347

[5] Naseh M, Heydari H. Analytical method for levitation force calculation of radial HTS magnetic bearings. IET Electric Power Applications. 2017;**11**: 369-377. DOI: 10.1049/iet-epa.2016.0070

[6] Kim SB, Ikegami T, Fujii Y, Takahashi M, Onodera H. Development of the noncontact rotating system using combined ring-shaped HTS bulks and permanent magnets. IEEE Transactions on Applied Superconductivity. 2014;**24**: 6800105. DOI: 10.1109/TASC.2013.2284021

[7] Arsenio AJ, Roque M, Cardeira C, Costa Branco PJ, Melicio R. Prototype of a zero-field-cooled YBCO bearing with continuous ring permanent magnets. IEEE Transactions on Applied Superconductivity. 2018;**28**:5207407. DOI: 10.1109/tasc.2018.2817279

[8] Mukoyama S et al. Development of superconducting magnetic bearing for 300 kW flywheel energy storage system. IEEE Transactions on Applied Superconductivity. 2017;**27**:3600804. DOI: 10.1109/ TASC.2017.2652327

[9] Wang J, Wang S. High Temperature Superconducting Magnetic Levitation. Berlin, Boston: De Gruyter; 2017. p. 388. DOI: 10.1515/9783110538434

[10] Floegel-Delor U, Schirrmeister P, Riedel T, Koenig R, Kantarba V, Werfel FN. Bulk superconductor levitation devices: Advances in and prospects for development. IEEE Transactions on Applied Superconductivity. 2018;**28**:3601605. DOI: 10.1109/tasc.2018.2809467

[11] Sotelo G, Dias D, Andrade R, Stephan R. Tests on a superconductor linear magnetic bearing of a full-scale MagLev vehicle. IEEE Transactions on Applied Superconductivity. 2011;**21**: 1464-1468. DOI: 10.1109/TASC.2010.2086034

[12] Kurbatov P, Kurbatova E, Dergachev P, Kulayev Y. Simulation of the body motion in a tube with the linear HTS suspension. In: 2018 IEEE 18th International Power Electronics and Motion Control Conference (PEMC-2018); 26–30 August 2018; Budapest, Hungary. New York: IEEE; 2018. pp. 611-616. DOI: 10.1109/EPEPEMC.2018.8521935

[13] Kovalev K, Koneev S, Poltavec V, Gawalek W. Magnetically levitated high-speed carriages on the basis of bulk HTS elements. In: Proceedings of 8th International Symposium on Maglev Suspension Technology (ISMST'8), 26–28 September 2005; Dresden, Germany. 2005. p. 51

[14] Deng Z et al. A high-temperature superconducting Maglev ring test line developed in Chengdu, China. IEEE Transactions on Applied Superconductivity. 2016;**26**:3602408. DOI: 10.1109/TASC.2016.2555921

[15] Bean C. Magnetization of high-field superconductors. Reviews of Modern Physics. 1964;**36**:31-39. DOI: 10.1103/RevModPhys.36.31

[16] Kim Y, Heampstead C, Strnad A. Critical persistent currents in hard superconductors. Physical Review Letters. 1962;**9**:306-309. DOI: 10.1103/PhysRevLett.9.306

[17] Ruiz-Alonso D, Coombs T, Campbell A. Numerical analysis of high-temperature superconductors with the critical state model. IEEE Transactions on Applied Superconductivity. 2004;**14**: 2053-2063. DOI: 10.1109/TASC.2004.838316

[18] Chun Y et al. Finite element analysis of magnetic field in high temperature bulk superconductor. IEEE Transactions on Applied Superconductivity. 2001;**11**: 2000-2003. DOI: 10.1109/77.920246

[19] Bean C. Response of high temperature superconductors to a step in magnetic field. In: Superconductivity and Applications (Proceedings of the Third Annual Conference on Superconductivity and Applications): 19–21 September 1989; New York. New York: Springer Science; 1989. pp. 767-772

[20] Sinder M, Meerovich V, Sokolovsky V, Vajda I. Penetration of magnetic field into high-temperature superconductors. IEEE Transactions on Applied Superconductivity. 1999;**9**: 4661-4665. DOI: 10.1109/77.819335

[21] Kim Y, Hempstead C, Strand A. Flux-flow resistance in type-II superconductors. Physical Review. 1965; **139**:A1163-A1172. DOI: 10.1103/PhysRev.139.A1163

[22] Vinen W, Warren A. Flux flow resistivity in type I superconductors: II. Theoretical discussion. Proceedings of the Physical Society. 1967;**91**:409-421. DOI: 10.1088/0370-1328/91/2/319

[23] Prigozhin L. Analysis of critical-state problems in type-II superconductivity. IEEE Transactions on Applied Superconductivity. 1997;7: 3866-3873. DOI: 10.1109/77.659440

[24] Rhyner J. Magnetic properties and AC-losses of superconductors with power-law current-voltage characteristics. Physica C. 1993;**212**: 292-300. DOI: 10.1016/0921-4534(93)90592-E

[25] Zou S, Zermeño V, Grilli F. Influence of parameters on the simulation of HTS bulks magnetized by pulsed field magnetization. IEEE Transactions on Applied Superconductivity. 2016;**26**:4702405. DOI: 10.1109/TASC.2016.2535379

[26] Douine B, Bonnard C, Sirois F, Berger K, Kameni A, Lévêque J. Determination of J_C and n-value of HTS pellets by measurement and simulation of magnetic field penetration. IEEE Transactions on Applied Superconductivity. 2015;**25**:8001008. DOI: 10.1109/TASC.2015.2409201

[27] Yokono T, Hasegawa K, Kamitani A. Magnetic shielding analysis of high-Tc superconducting plates by power law, flux-flow, and flux-creep models. IEEE Transactions on Applied Superconductivity. 2003;**13**:1672-1675. DOI: 10.1109/TASC.2003.812860

[28] Ikuno S, Kamitani A. Shielding current density analysis of axisymmetric HTS by element-free galerkin method. IEEE Transactions on Applied Superconductivity. 2005;**15**:3688-3691. DOI: 10.1109/TASC.2005.849393

[29] Anderson P. Theory of flux creep in hard superconductors. Physical Review Letters. 1962;**9**:309-311. DOI: 10.1103/PhysRevLett.9.309

[30] Klimenko E, Imenitov A, Shavkin S, Volkov P. Resistance–current curves of high pinning superconductors. Journal of Experimental and Theoretical Physics. 2005;**100**:50-65. DOI: 10.1134/1.1866198

[31] Kulaev Y et al. Modeling of electrophysical properties of bulk high-temperature superconductors in calculations of magnetic systems. Russian Electrical Engineering. 2015;**86**:213-219. DOI: 10.3103/S1068371215040070

[32] Stratton J. Electromagnetic Theory. New York: McGraw-Hill; 1941. p. 631

[33] Kulaev Y, Kurbatov P, Kurbatova E. Construction of combined models of the properties of bulk high-temperature superconducting materials. Russian Electrical Engineering. 2017;**88**:465-470. DOI: 10.3103/s1068371217070094

Chapter 4

Generalized BCS Equations: A Review and a Detailed Study of the Superconducting Features of $Ba_2Sr_2CaCu_2O_8$

Gulshan Prakash Malik and Vijaya Shankar Varma

Abstract

High-T_c superconductors (SCs) are most widely studied via the multiband approach (MBA) based on the work of Suhl et al. and the Nambu-Eliashberg-McMillan extension of the BCS theory. Complementing MBA and presented in a recent monograph is an approach based on the generalized BCS equations (GBCSEs), which too has been applied to a significant number of SCs. GBCSEs are obtained via a Bethe-Salpeter equation and the Matsubara technique. One of the key features of this approach is the characterization of a composite SC by Cooper pairs with different binding energies—each of which is identified with a gap (Δ) of the SC—depending on whether pairing is due to phonon exchanges with one, two, or more ion species. Another feature is the incorporation of chemical potential in the GBCSEs, which enables one to calculate the critical current density j_0 of an SC using the same parameters that determine its T_c and Δ. Following a review of the concepts of this approach, given herein, for the first time, is a detailed explanation of the multitude of reported empirical values of $\{T_c, \Delta, j_0\}$ of $Bi_2Sr_2CaCu_2O_8$. Also discussed are the currently topical issues of $s^{\pm}$-superconductivity and the isotope-like effect for high-T_c SCs.

Keywords: high-T_c superconductors, multiband approach, GBCSE-based approach, superconducting features of $Bi_2Sr_2CaCu_2O_8$, $s^{\pm}$-superconductivity, isotope-like effect for high-T_c superconductors, suggestions for further increasing the T_c of the Bi-based superconductors

1. Introduction

1.1 A historical note

The phenomenon of superconductivity was first observed by Kamerlingh Onnes [1] in 1911 while studying the electrical resistivity of metallic mercury as a function of temperature when cooled to liquid helium temperatures. He noticed that at the critical temperature T_c = 4.2 K (−269°C), the resistance vanished abruptly. This was an exciting discovery because it suggested that such materials could have immense practical applications. Since cooling a material to very low temperatures is a tedious and expensive process, Onne's discovery triggered a search in scientific laboratories across the world for other superconducting materials having T_cs higher than that of Hg, the ultimate aim being to find a material which is superconducting at room temperature. While this search led to the discovery of many elements and alloys that are superconductors (SCs), till as late as in the 1980s, the highest T_c of any known SC was about 23 K (for Nb_3Sn). A radical change occurred in this situation in 1986 when Bednorz and Müller announced the discovery of an SC (Ba-La-Cu-O system) with $T_c \approx 38$ K, for which they were awarded the Nobel Prize in 1987. An avalanche of activity followed soon thereafter leading to the discovery of many "high-temperature SCs" such as MgB_2 ($T_c \simeq 40$ K), Fe-based SCs (highest $T_c \simeq 55$ K), and "cuprates," examples of which are Bi-, Tl-, and Hg-based SCs that contain one or more units of CuO_2 as a constituent, the T_cs of these SCs at ambient pressure being about 95, 110, and 138 K, respectively, and for the Hg-based SCs, going up to 164 K at high pressure. The confirmed current record for the highest T_c (203 K; at a pressure of 150 GPa) is held by H_3S which was discovered in 2015. It would thus appear that realization of the dream of room temperature SCs is near at hand.

On the theoretical front, a clear explanation of superconductivity evaded the efforts of many theoreticians who had been engaged in the quest. It was not until 46 years had elapsed since Onnes' discovery that John Bardeen, Leon Cooper, and Robert Schrieffer (BCS) [2] were able to provide a generally accepted explanation of it for elemental SCs, for which they were awarded the Nobel Prize in 1972. Indeed, a contributing factor for this lag between theory and experiment was the tragedy of the two world wars in the intervening period.

BCS theory explains superconductivity as due to the formation of Cooper pairs (CPs), each of which is a bound state of two electrons. A well-known explanation for the origin of the attractive interaction causing electrons to be bound together is as follows: the passage of an electron through the ion lattice of an element leaves behind a deformation trail of enhanced positive charges. If a second electron passes through the lattice while it is recovering from the effect of the passage of the first, it will experience an attractive force that may be greater than the Coulomb repulsion between them. Hence, overall, the electron-lattice-electron interaction can be a net *weak* attractive interaction that leads to the formation of CPs. Since the bound state of two particles has lower energy than the state in which they are both free, the former becomes the preferred state leading to the occurrence of superconductivity. In the language of quantum field theory, it is said that the pairing of electrons takes place because they exchange a phonon due to the effect of the ion lattice. BCS theory is hence a weak-coupling theory which works for elemental SCs because the highest T_c for this class of SCs is about 9 K (for Nb). A general perception about this theory is that it is inadequate to explain the occurrence of such high-T_cs as have been observed. This is indeed so if the one-phonon exchange mechanism (1PEM) is considered to be the sole mechanism responsible for pairing. Such a view overlooks the fact that *all* high-T_c SCs are composite materials consisting of sub-lattices of more than one ion species and therefore the theory ought to be generalized to address the situation where CPs may also be formed due to more than 1PEM. Interestingly, as noted below, such a generalization can be carried out based on three theoretical developments that took place in the same decade, 1950s, to which the BCS theory belongs.

1.2 GBCSEs and their conceptual basis

Two of the three developments alluded to the above are widely known: the invention of the Bethe-Salpeter equation (BSE) in 1951 [3] and, in analogy with quantum field theory, the creation of a finite-temperature field theory (FTFT) by

Matsubara [4] in 1955. Not so well-known is the third development that took place in 1954 due to Okubo [5], viz., the introduction of the concept of a superpropagator as the propagator for the non-polynomial field exp ($g\phi$) where g is a coupling constant and ϕ a scalar field.

BSE is a four-dimensional equation which was invented for the description of bound states of two relativistic particles, hordes of which were being discovered in the late 1940s and early 1950s in cosmic rays and high-energy accelerators around the world. A general perception about the BSE is that it is irrelevant for superconductivity which is a strictly nonrelativistic phenomenon.

FTFT is obtained via an adaption of quantum field theory where time is replaced by temperature as the variable in terms of which evolution of a system is studied. Because temperature is a statistical concept, this is a huge step which converts an equation setup for the bound states of two particles interacting in vacuum into an equation valid in a many-body system. A shortcut known as the Matsubara prescription or recipe for introducing temperature into a BSE is to replace in it the component q_4 of the four-vector $q_\mu = (q_4, \mathbf{q})$ as follows (e.g., [6]):

$$\begin{aligned} q_4 &= (2n+1)\pi/(-i\beta) \quad \text{for fermions} \\ &= 2\text{n}\pi/(-i\beta) \qquad \text{for bosons} \\ \int dq_4 &= \frac{2\pi}{-i\beta}\sum_{n=-\infty}^{\infty} \end{aligned} \tag{1}$$

Okubo's work remained an obscure academic exercise till Salam and collaborators found that non-polynomial theories have inbuilt damping effects. It was therefore hoped that they might provide the means of renormalizing, e.g., the weak interactions. These theories were not pursued any further after the gauge theory of interactions became widely accepted as the correct theory. However, it is relevant in the context of superconductivity to note that the expression for the Feynman superpropagator corresponding to the non-polynomial interaction $1/(1+g\phi)$ is [7]:

$$\Delta_F^{\text{SP}}(k) \approx \frac{1}{k^2g^2} - \int_0^\infty dt \frac{e^{-t}(2-t)}{k^2g^2+t}. \tag{2}$$

The comparison of the above with the expression for one-particle Feynman propagator, viz.,

$$\Delta_F^{1-\text{Particle}}(k) = \frac{1}{k^2 - i\varepsilon} \tag{3}$$

reveals that a superpropagator represents a weighted superposition of multiple quanta. The significance of this remark emerges if we recall how the electron-lattice-electron interaction in an elemental SC causes formation of CPs via 1PEM. Appealing to the same picture for a composite SC, it follows that pairing can now also be caused via more than 1PEM because sub-lattices of ions of different masses are affected differently by the passage of electrons through the lattice of the SC. We are hence led to the following outline of a novel strategy for dealing with multicomponent SCs: start with a BSE for the bound states of two fermions, temperature-generalize it via the Matsubara prescription, and, for the kernel of the equation, employ a superpropagator representing exchange of multiple phonons for pairing.

To implement the above program, we need to employ (i) the instantaneous and mean-field approximations (IA, MFA), (ii) a model for fixing the Debye temperatures of all the ions that may cause pairing, and (iii) the Bogoliubov constraint on the interaction parameters obtained via solutions of the equations that the BSE leads to when multi-phonon exchange mechanisms are operative. An account of these concepts is as follows.

By IA is meant that we ignore the time of propagation of the quanta, the exchange of which causes the electrons to be bound together. For example, in the momentum-space transform of the one-particle propagator in Eq. (3) where $k^2 = k_0^2 - \boldsymbol{k}^2, k_0^2$ is equated to zero. This is an essential step for the application of the Matsubara prescription which introduces temperature into the theory and reduces the BSE to a three-dimensional equation which is then subjected to the nonrelativistic approximation appropriate for superconductivity.

Employing MFA in the 1PEM scenario, one approximates the propagator noted in Eq. (3) by a constant, $-V$ (as in [N(0)V], $\hbar = c = 1$), because pairing takes place in a very small region demarcated by $(E_F - k\theta \leq p^2/2m \leq E_F + k\theta)$, where E_F denotes the Fermi energy, $p^2/2m$ the energy of an electron, and θ the Debye temperature of the SC; outside of this region, V = 0. Similarly, in the nPEM scenario, in lieu of the expression for the superpropagator in Eq. (2), one employs the expression $- [V_1 + V_2 + \ldots + V_n]$, where n is the number of ion species that cause the electrons to form a pair, V_i corresponds to the phonon exchanged due to the ith species of ions of which the Debye temperature is θ_i, and $(E_F - k\theta_i \leq p^2/2m \leq E_F + k\theta_i)$ is the region of pairing due to these ions, outside of which $V_i = 0$.

Given the Debye temperature θ of any composite SC, we now need to fix the Debye temperatures of its constituent ion species, θ_1, θ_2, etc., which in general *must* be different from θ because, as was first pointed out by Born and Karman a long time ago, elastic waves in an anisotropic material travel with different velocities in different directions; for a detailed account of their work, see [8]. We now draw attention to the fact that sublattices of all the high-T_c SCs are composed of layers that contain either one or predominantly two ion species. The assumption that θ is also the Debye temperature of each of these sub-lattices then fixes the Debye temperature of the sub-lattice of layers that contains a single ion species as θ. As concerns fixing θ_1 and θ_2 for the sub-lattice comprising two ion species, we note that the following relation has been frequently used for finding the Debye temperature θ of a binary A_xB_{1-x} when the Debye temperatures θ_A and θ_B are known:

$$\theta = x\theta_A + (1 - x)\theta_B. \tag{4}$$

Since we are faced with the inverse of the above situation, i.e., given θ, we have to fix θ_A and θ_B, we now need to supplement Eq. (4) by another equation. We meet this requirement by assuming that the modes of vibration of the A and B ions simulate the weakly coupled modes of vibration of a double pendulum. This is suggested by the polariton effect where coupling between the phonon and the photon modes leads to two new modes, different from the original modes of either of them. With A as the upper bob in the double pendulum, we have our second equation as [10].

$$\frac{\theta_A}{\theta_B} = \left[\frac{1 + \sqrt{m_B/(m_A + m_B)}}{1 - \sqrt{m_B/(m_A + m_B)}}\right]^{1/2}, \tag{5}$$

where m_A and m_B are the atomic masses of the ions.

Given θ ($Bi_2Sr_2CaCu_2O_8$) = 237 K, m_{Bi} = 208.98, m_{Sr} = 87.62, and m_O = 15.999 (amu), the application of Eqs. (4) and (5) leads to

$$\theta_{Ca} = 237\text{ K}, \theta_{Bi} = 269\text{ K}, \theta_{Sr} = 286\text{ K}. \tag{6}$$

We note that (i) θ_{Ca} equals the Debye temperature of the SC because the layers containing Ca ions do not have any other constituent. (ii) The value of θ_{Bi} corresponds to Bi being the upper bob of the double pendulum and O the lower bob in the BiO layers; the other value of θ_B when Bi is the lower bob is 57 K. (iii) The value of θ_{Sr} also corresponds to Sr. being the upper bob of the double pendulum in the SrO layers; the other value of θ_{Sr} is 81 K. (iv) The values of θ_O in the two layers are not noted because they do not play a direct role in the pairing process. (v) Since we have four distinct choices for values of the pair $\{\theta_{Bi}, \theta_{Sr}\}$, we need a criterion for choosing one among them, and this is provided by the Bogoliubov constraint discussed below.

In a rigorous study concerned with renormalization of the BCS theory, starting with the assumption of a net attractive interaction between electrons, Bogoliubov found that the interaction parameter λ governing the pairing process must not exceed 0.5, otherwise the system would be unstable; for a detailed discussion, see [9]. Imposition of this constraint on the λs obtained via GBCSEs for high-T_c SCs is crucial. This is so because if one calculates λs for each of the four choices of θs as noted above, one can straightway reject those that lead to any of them that are negative and choose among the rest the ones that lead to values closest to 0.5. The values of θs in Eq. (6) have been chosen on this basis.

1.3 Plan of the chapter

This chapter is organized as follows. In Section 2.1 is given the parent Bethe-Salpeter equation from which GBCSEs for the T_c and Δs of a multicomponent SC are derived. In Section 2.2, after an outline of the key steps of their derivation, are given the GBCSEs which are constrained by the inequality $E_F \gg k\theta$. More generally, E_F-incorporated GBCSEs, which are not constrained by this inequality, are given in Section 2.3. This is followed up by an account of similar equations, i.e., subject to and not subject to the said inequality, for the dimensionless construct $y = (k\theta/P_0)\sqrt{2m/E_F}$, which is required for calculating j_0 and several other parameters of an SC. In Section 4 are listed the properties of Bi-2212, the explanation of which is taken up in Section 5. In Section 6 we draw attention to the application of the framework of GBCSEs to a diverse variety of SCs, viz., LCO, the heavy fermion, and the Fe-based SCs. Also included in this section is a brief account of a recent study concerned with the isotope-like effect for Bi-2212. Sections 6 and 7 are devoted, respectively, to the discussion of the key features of our approach and the conclusions following from them.

2. The framework of GBCSEs

After an outline of the key steps of their derivation, we give in this section the GBCSEs that are later employed for the study of various superconducting features of $Bi_2Sr_2CaCu_2O_8$ (Bi-2212). We refer the reader to [10] for a detailed derivation of these equations.

2.1 The parent BSE

The parent equation from which GBCSEs are derived is the following relativistic BSE [11] for the bound states of fermions 1 and 2 of equal mass (m):

$$\psi\left(p_\mu\right) = (-2\pi i)^{-1} \int d^4 q_\mu \frac{1}{\gamma_\mu^{(1)} P_\mu/2 + \gamma_\mu^{(1)} q_\mu - m + i\varepsilon} \\ \text{x} \frac{1}{\gamma_\mu^{(2)} P_\mu/2 + \gamma_\mu^{(2)} q_\mu - m + i\varepsilon} I\left(q_\mu - p_\mu\right)\psi\left(q_\mu\right), \tag{7}$$

where ψ is the BS amplitude for the formation of the bound state of the two particles, P_μ the total four-momentum of their center of mass, q_μ their relative four-momentum, $\gamma_\mu^{(1,2)}$ are their Dirac matrices, and $I(q_\mu - p_\mu)$ is the kernel of the equation that causes the particles to be bound together; the metric employed is time-preferred: $a_\mu b_\mu = a_4 b_4 - \mathbf{a.b}$.

2.2 GBCSEs corresponding to multiphonon exchanges for pairing

We now outline how GBCSEs are derived when CPs in an SC are formed due to multi-phonon exchange mechanisms subject to the following constraint:

$$E_F >> k\theta, \tag{8}$$

where θ is the Debye temperature of the SC.
Beginning with Eq. (7), we obtain GBCSEs by:

1. Employing IA: $I\left(q_\mu - p_\mu\right) = I(\boldsymbol{q} - \boldsymbol{p})$

2. Carrying out spin reduction of the equation by multiplying it with Dirac matrices $\gamma_4^{(1)}\gamma_4^{(2)}$

3. Working in the rest frame of the two particles: $P_\mu = (E, 0)$

4. Putting $E = 2E_F + W$

5. Temperature-generalizing the equation via the Matsubara recipe given in the previous section

6. Assuming that the signature of W changes on crossing the Fermi surface

7. Choosing for the kernel a generalized form of the model BCS interaction:

$$I(\boldsymbol{q} - \boldsymbol{p}) = \sum_{i=1}^{n} \frac{-V_i}{(2\pi)^3} \left(\text{for } E_F - k\theta_i \leq \frac{\boldsymbol{p}^2}{2m}, \frac{\boldsymbol{q}^2}{2m} \leq E_F + k\theta_i\right) \\ = 0 \text{ (otherwise)}, \tag{9}$$

where n is the number of ion species due to which pairs are formed and θ_i is the Debye temperature of the ith species of ions

8. Calculating θi via the Debye temperature θ of the SC and Eqs. (4) and (5) given above

With their application to Bi-2212 in mind, given below are the equations that the above steps lead to in the scenario in which the smaller of the two gaps of an SC is due to a two-phonon exchange mechanism (2PEM) and its larger gap and the highest T_c are due to a three-phonon exchange mechanism (3PEM) (reminder: $E_F >> k\theta$)

$$1 = \lambda_1 \ln\left[1 + \frac{2k\theta_1}{|W_2|}\right] + \lambda_2 \ln\left[1 + \frac{2k\theta_2}{|W_2|}\right] \tag{10}$$

$$1 = \lambda_1 \ln\left[1 + \frac{2k\theta_1}{|W_3|}\right] + \lambda_2 \ln\left[1 + \frac{2k\theta_2}{|W_3|}\right] + \lambda_3 \ln\left[1 + \frac{2k\theta_3}{|W_3|}\right] \tag{11}$$

$$1 = \lambda_1 \int_0^{\theta_1/2kT_c} dx \frac{\tanh(x)}{x} + \lambda_2 \int_0^{\theta_2/2kT_c} dx \frac{\tanh(x)}{x} + \lambda_3 \int_0^{\theta_3/2kT_c} dx \frac{\tanh(x)}{x}, \tag{12}$$

where $\lambda_i = [N(0)V_i], N(0) = (2m)^{3/2}E_F^{1/2}/4\pi^2$ $(\hbar = c = 1)$.

Remark: If $\lambda_2 = 0$ in Eq. (10) and $\lambda_2 = \lambda_3 = 0$ in both Eqs. (11) and (12), then we have a one-phonon exchange mechanism (1PEM) in operation. In this case Eq. (12) becomes identical with the BCS equation for the T_c of an elemental SC, whereas Eq. (10), or Eq. (11), leads to

$$|W_1| = 2k\theta/[\exp(1/\lambda) - 1].$$

Recalling the BCS equation for the gap of an elemental SC at $T = 0$, i.e.,

$$\Delta_0 = k\theta/\sinh(1/\lambda),$$

we observe that $|W_1| = \Delta_0 = 2k\theta \exp(-1/\lambda)$, when $\lambda \to 0$. This per se suggests that one may employ the equation for $|W_1|$ in lieu of the equation for Δ_0. Based on the values of $|W_1|$ and Δ_0 obtained via a detailed study of six elemental SCs [10] by employing the empirical values of their Debye temperatures, it has been shown that $|W_1|$ is a viable alternative to Δ_0. We are thus led to assume that each of the three Ws following from Eq. (11) is to be identified with a gap of the SC and that $|W_3|>|W_2|>|W_1|$, where $|W_2|$ corresponds to the solution of the equation when one of the λs is zero and $|W_1|$ to when two of them are zero.

2.3 E_F-incorporated GBCSEs

It has been known for a long time that T_cs of some SCs—$SrTiO_3$ being a prime example—depend in a marked manner on their electron concentration n_s which is related to E_F. Recent studies—both experimental [12–14] and theoretical [15–17]—too suggest that *low* values of E_F play a fundamental role in determining the properties of high-T_c SCs. This makes it natural to seek equations for the T_cs and Δs of these SCs that incorporate E_F as a variable and are therefore not constrained by inequality (8). Such equations are in fact an integral part of the BCS theory. Their application to dilute $SrTiO_3$ by Eagles [18] perhaps marks the beginning of what is now known as BCS-BEC crossover physics. It seems to us that the focus of crossover physics has, by and large, been different from that of the studies of high-T_c SCs.

Beginning with Eq. (7) and carrying out steps (1–8) noted above, without imposing inequality (8), the μ-incorporated GBCSEs corresponding to Eqs. (10)–(12) are as follows [10]:

$$\frac{\lambda_1}{2}a_1(\mu_0) + \frac{\lambda_2}{2}a_2(\mu_0) = \left[\frac{3}{4}a_3(\mu_0)\right]^{1/3} \tag{13}$$

$$\frac{\lambda_1}{2}b_1(\mu_0) + \frac{\lambda_2}{2}b_2(\mu_0) + \frac{\lambda_3}{2}b_3(\mu_0) = \left[\frac{3}{4}b_4(\mu_0)\right]^{1/3} \tag{14}$$

$$\frac{\lambda_1}{2}c_1(\mu_1) + \frac{\lambda_2}{2}c_2(\mu_1) + \frac{\lambda_3}{2}c_3(\mu_1) = \left[\frac{3}{4}c_4(\mu_1)\right]^{1/3}, \tag{15}$$

where μ_0 (μ_1) is the chemical potential at T = 0 (T = T_c) and μ_0 has been used interchangeably with E_F,

$$a_1(\mu_0) = \boldsymbol{Re}\left[\int_{-k\theta_1}^{k\theta_1} d\xi \frac{\sqrt{\xi+\mu_0}}{|\xi| + |W_2|/2}\right], a_2(\mu_0) = \boldsymbol{Re}\left[\int_{-k\theta_2}^{k\theta_2} d\xi \frac{\sqrt{\xi+\mu_0}}{|\xi| + |W_2|/2}\right]$$

$$a_3(\mu_0) = \boldsymbol{Re}\left[\frac{4}{3}(\mu_0 - k\theta_m)^{3/2} + \int_{-k\theta_m}^{k\theta_m} d\xi\sqrt{\xi+\mu_0}(1 - \frac{\xi}{\sqrt{\xi^2 + W_2^2}}\right]$$

$$b_1(\mu_0) = \boldsymbol{Re}\left[\int_{-k\theta_1}^{k\theta_1} d\xi \frac{\sqrt{\xi+\mu_0}}{|\xi| + |W_3|/2}\right], b_2(\mu_0) = \boldsymbol{Re}\left[\int_{-k\theta_2}^{k\theta_2} d\xi \frac{\sqrt{\xi+\mu_0}}{|\xi| + |W_3|/2}\right]$$

$$b_3(\mu_0) = \boldsymbol{Re}\left[\int_{-k\theta_3}^{k\theta_3} d\xi \frac{\sqrt{\xi+\mu_0}}{|\xi| + |W_3|/2}\right]$$

$$b_4(\mu_0) = \boldsymbol{Re}\left[\frac{4}{3}(\mu_0 - k\theta_2)^{3/2} + \int_{-k\theta_m}^{k\theta_m} d\xi\sqrt{\xi+\mu_0}\left(1 - \frac{\xi}{\sqrt{\xi^2 + W_3^2}}\right]\right.$$

$$c_1(\mu_1) = \boldsymbol{Re}\left[\int_{-\theta_1/2T_{c3}}^{\theta_1/2T_{c3}} dx\sqrt{2kT_{c3}x+\mu_1}\frac{\tanh(x)}{x}\right],$$

$$c_2(\mu_1) = \boldsymbol{Re}\left[\int_{-\theta_2/2T_{c3}}^{\theta_2/2T_{c3}} dx\sqrt{2kT_{c3}x+\mu_1}\frac{\tanh(x)}{x}\right]$$

$$c_3(\mu_1) = \boldsymbol{Re}\left[\int_{-\theta_3/2T_{c3}}^{\theta_3/2T_{c3}} dx\sqrt{2kT_{c3}x+\mu_1}\frac{\tanh(x)}{x}\right]$$

$$c_4(\mu_1) = \boldsymbol{Re}\left[2kT_{c3}\int_{-\theta_m/2T_{c3}}^{\theta_m/2T_{c3}} dx\sqrt{2kT_{c3}x+\mu_1}\{1 - \tanh(x)\}\right],$$

θ_m is the Debye temperature that has the highest value among all the Debye temperatures being used (when $\lambda_2 \neq 0, \theta_m = \theta_2$; when $\lambda_2 = 0, \theta_m = \theta_3$; when $\lambda_2 = \lambda_3 = 0, \theta_m = \theta_1$), and the operator $\boldsymbol{Re}$ ensures that the integrals yield real values even when expressions under the radical signs are negative.

2.4 Equation for $y = (k\theta/P_0)\sqrt{2m/E_F}$ (E_F>>$k\theta$)

If it turns out that we need to employ the μ-incorporated Eqs. (13)–(15) to deal with high-T_c SCs, rather than Eqs. (10)–(12), as will be shown to be the case, then there will be a need for at least one more equation for an additional property of the SC, besides equations for one of its T_c and Δs, in order to fix the values of μs and λs that lead to various empirical features of the SC. We choose the critical current density j_0 of the SC at T = 0 to meet this need.

The BSE-based approach enables one to derive an equation for j_0 of an SC by employing the same values of θs and λs that determine a given value of its T_c and different values of its Δs. This comes about because one can now obtain expressions for the effective mass m* of an electron, critical velocity v_0 at T = 0, and the number of superconducting electrons n_s at T = 0, and hence for j_0 in terms of y and the values of the following other parameters of the SC: θ, the electronic specific heat constant $\gamma, s \equiv m^*/m_e$, v_0, n_s, and v_g, where m_e is the free electron mass and v_g the gram-atomic volume of the SC.

As concerns y, we recall that in the original BCS theory, the Hamiltonian was restricted at the outset to comprise terms corresponding to pairs having zero center-of-mass momentum. The BSE-based approach enables one to go beyond this restriction by setting $P_\mu = (E, \mathbf{P})$, rather than $(E, 0)$, where **P** is the three-momentum of CPs in the lab frame. Beginning with Eq. (7) and carrying out essentially the same steps as for the $P_\mu = (E, 0)$ case, we now obtain in the 3PEM scenario the following equation for y [10], which can be solved after the λs have been determined by solving Eqs. (10)–(12):

$$1 = \lambda_1\, t(r_1, y) + \lambda_2\, t(r_2, y) + \lambda_3\, t(r_3, y), \tag{16}$$

where

$$t(r, y) = \left[r\, y \ln\left(\frac{r\, y}{r\, y - 1}\right) + \ln\left(r\, y - 1\right) \right], \ \mathrm{r}_i = \theta_i/\theta.$$

Note that equating one (two) of the λs in Eq. (16) to zero, we obtain the equation for y in the 2PEM (1PEM) scenario.

In terms of y and the other parameters mentioned above, it has been shown in [19, 20] that

$$s \equiv m^*/m_e = A_1 \frac{\left(\gamma/v_g\right)^{2/3}}{E_F^{1/3}} \tag{17}$$

$$n_s(E_F) = A_2 \left(\gamma/v_g\right) E_F \tag{18}$$

$$P_0(E_F) = A_3 \frac{\theta}{y} \frac{\left(\gamma/v_g\right)^{1/3}}{E_F^{2/3}} \tag{19}$$

$$v_0(E_F) = A_4 \frac{\theta}{y} \frac{1}{\left(\gamma/v_g\right)^{1/3} E_F^{1/3}} \tag{20}$$

and hence

$$j_0(E_F) = \frac{n_s(E_F)}{2} e^* v_0(E_F) = A_5 \frac{\theta}{y} \left(\gamma/v_g\right)^{2/3} E_F^{2/3}, \quad (e^* = 2e) \tag{21}$$

where e is the electronic charge, and
$A_1 \cong 3.305x10^{-10}\ \mathrm{eV}^{-1/3}\mathrm{cm}^2\mathrm{K}^{4/3}, A_2 \cong 2.729x10^7\ \mathrm{eV}^{-2}\mathrm{K}^2$,
$A_3 \cong 1.584x10^{-6}\ \mathrm{cm\ K}^{-1/3}$, $A_4 \cong 1.406x10^8\ \mathrm{eV}^{2/3}\ \mathrm{sec}^{-1}\mathrm{K}^{-5/3}$, and
$A_5 \cong 6.146x10^{-4}\ \mathrm{CeV}^{-4/3}\mathrm{K}^{1/3}\mathrm{s}^{-1}$.

2.5 E_F-dependent equation for $y = (k\theta/P_0)\sqrt{2m/E_F}$

If the T_c and Δs of an SC are studied via Eqs. (10)–(12), which are valid when $E_F >> k\theta$, then it is appropriate to employ Eq. (16) for y because it too is valid when the same inequality is satisfied. If it turns out that we need to employ the μ-incorporated Eqs. (13)–(15) in lieu of Eqs. (10)–(12), then consistency demands that we employ for y, in lieu of Eq. (16), an equation that too is explicitly μ-dependent.

The desired equation for y in the 3PEM scenario, obtained by setting $P_\mu = (E, \boldsymbol{P})$ in the BSE and doing away with the earlier constraint on E_F, is [21]:

$$1 = \frac{\lambda_1}{4} T(\Sigma_1', r_1 y) + \frac{\lambda_2}{4} T(, \Sigma_2', r_2 y) + \frac{\lambda_3}{4} T(\Sigma_3', r_3 y), \tag{22}$$

where

$$T(\Sigma', y) = \boldsymbol{Re}\left[\int_0^1 dx F(\Sigma', x, y)\right], F(\Sigma', x, y) = f_1(\Sigma', x, y) + f_2(\Sigma', x, y)$$

$$f_1(\Sigma', x, y) = 4\left[-u_1(\Sigma', x, y) - u_2(\Sigma', x, y) + u_3(\Sigma', x, y) + u_4(\Sigma', x, y)\right],$$

$$f_2(\Sigma', x, y) = 2\ln\left[\frac{1+u_1(\Sigma', x, y)}{1-u_1(\Sigma', x, y)} \frac{1+u_2(\Sigma', x, y)}{1-u_1(\Sigma', x, y)} \frac{1-u_3(\Sigma', x, y)}{1+u_3(\Sigma', x, y)} \frac{1-u_4(\Sigma', x, y)}{1+u_4(\Sigma', x, y)}\right]$$

$$u_1(\Sigma', x, y) = \sqrt{1 - \Sigma' x/y}, u_2(\Sigma', x, y) = \sqrt{1 + \Sigma' x/y}$$

$$u_3(\Sigma', x, y) = \sqrt{1 - \Sigma'(1 - x/y)}, u_4(\Sigma', x, y) = \sqrt{1 + \Sigma'(1 - x/y)}$$

$$\Sigma' = k\theta/E_F, \Sigma_i' = k\theta_i/E_F, \quad r_i = \theta_i/\theta.$$

Equating one (two) of the λs in Eq. (22) to zero, we obtain the μ-dependent equation for y in the 2PEM (1PEM) scenario.

3. Superconducting features of $Bi_2Sr_2CaCu_2O_8$ addressed in this paper

The reported empirical values of {T_c, Δ} of the above SC are [22], p 447:

$$\{T_c\ (\mathrm{K}), \Delta\ (\mathrm{meV})\} = \{95, 38\}, \{86, 28\}, \{62, 18\}. \tag{23}$$

As concerns empirical values of j_0 of the SC, the situation is rather complicated because 14 values of this parameter have been reported in [22], p 468—depending upon the shape of the SC, how it is doped, and the values of temperature and the external magnetic field H under which they were determined. However, none of these values corresponds to the T = H = 0 scenario employed by us. The value $j_0 = 10^6$ A/cm^2 at T = 4.4 K and unspecified value of H being closest to it. Another value of interest for us is [23]:

$$j_0 = 3.2\mathrm{x}10^7\ \mathrm{A/cm^2}\ (T = 4.2\ \mathrm{K}, H = 12\ \mathrm{Tesla}). \tag{24}$$

In the following we give a quantitative explanation of the values of the parameters noted in Eqs. (23) and (24) and a plausible explanation of the multitude of the other values of these parameters.

4. An explanation of the multiple {T_c, Δ, j_0}-values of Bi-2212 via GBCSEs

4.1 Earlier work

In our earlier work dealing with the above SC [10], we had employed GBCSEs constrained by inequality (8) and had restricted ourselves to the 2PEM scenario. The maximum value of Δ that we could then account for, subject to the Bogoliubov constraint, was 20.4 meV corresponding to T_c = 95 K. Subsequently, even after employing μ-incorporated GBCSEs which are not constrained by inequality (8), we found that the 2PEM scenario was inadequate in explaining a value of Δ exceeding 20.4 meV [20, 21]. What remained elusive in these studies were especially the value Δ = 38 meV and the other values of T_c and Δ noted in Eq. (23). We show below how this lacuna is met via 3PEM in the present work.

4.2 GBCSEs constrained by inequality $E_F \gg k\theta$: premises and results

Since working in the 2PEM scenario, both with the μ-independent and μ-dependent GBCSEs, the maximum value of Δ satisfying the Bogoliubov constraint was found to be 20.4 meV; we assume here that among the values noted in Eq. (23),

$$\Delta_2 = 18 \text{ meV}, \Delta_3 = 38 \text{ meV}, T_{c3} = 95 \text{ K}, \tag{25}$$

i.e., as our notation suggests, we attribute the first parameter above to 2PEM and the other two parameters to 3PEM. With θs as given in Eq. (6), the solutions of Eqs. (10–12) lead to $\{\lambda_{Ca}, \lambda_{Sr}, \lambda_{Bi}\} = \{42.7, -37.6, 1.38\}$. Since these values are in conflict with the Bogoliubov constraint, in the absence of any other handle, we now systematically fine-tune the input values of the two Δs and T_c. We thus find that λs satisfying this constraint can be found only for values of Δs and T_{c3} that are substantially different from their empirical values in Eq. (23). An example, in order to obtain the set $\{\lambda_{Ca}, \lambda_{Sr}, \lambda_{Bi}\} = \{0.4652, 0.2291, 0.4856\}$ which satisfies the required constraint, we have to choose input values as $\Delta_2 = 14 meV$, $\Delta_3 = 33.5\ meV$, and $T_{c3} = 141$ K.This is an unacceptable state of affairs, suggesting the need to have a handle other than variation of the Δs- and T_c-values of the SC to satisfy the Bogoliubov constraint. We show below how the μ-incorporated Eqs. (13)–(15), which are not constrained by inequality (8), meet this need.

4.3 Interaction parameters obtained via μ-incorporated GBCSEs

Before attempting to find the three λs satisfying the Bogoliubov constraint by employing Eqs. (13)–(15) for the same inputs as in Eqs. (6) and (25) and using μ as a handle, as a consistency check of these equations, we solve them for a value of $\mu_0 \gg k\theta$, say, $\mu_0 = 100\ k\ \theta$. For this value of $\mu_0 = \mu_1$, or any other greater value, we find that $\lambda_{Ca} = 42.69$, $\lambda_{Sr} = -37.59$, and $\lambda_{Bi} = 1.38$, which are precisely the values we had obtained above via Eqs. (10)–(12) for the same inputs. This establishes that Eqs. (13)–(15) provide a reasonable generalization of Eqs. (10)–(12) and may therefore be used for any values of μ_0 and μ_1. If we now solve Eqs. (13)–(15) with the same inputs as before, but with progressively lower values of $\mu_0 = \mu_1$, we find that there is a marginal *increase* in the value of each of the λs; an example, for$\mu_0 = \mu_1 = 5\ k\ \theta$, $\lambda_{Ca} = 42.90$, $\lambda_{Sr} = -37.78$, $\lambda_{Bi} = 1.39$. This suggests that the assumption that $\mu_0 = \mu_1$ may not be valid. To check this, we solve our equations again with $\mu_0 = 5\ k\ \theta$ and $\mu_1 = 0.2\ \mu_0$, to find that $\lambda_{Ca} = 16.46$, $\lambda_{Sr} = -15.56$, and $\lambda_{Bi} = 0.1144$. It is thus confirmed that in adopting a low value of μ_0, with μ_1 a small fraction of it, we are proceeding in the right direction. Following this course, we obtain the desired set of λ values corresponding to the T_c and Δ values in Eq. (25) as

$$\{\lambda_{Ca}, \lambda_{Sr}, \lambda_{Bi}\} = \{0.3123, 0.4993, 0.5000\}, \tag{26}$$

when

$$\mu_0 = 1.95\ \text{k}\theta\ (39.82\ \text{meV}), \mu_1 = 0.105\ \mu_0\ (4.18\ \text{meV}). \tag{27}$$

4.4 Calculation of y via the μ-dependent Equation (22)

A consistency check of Eq. (22): While both the μ-independent Eq. (16) and the μ-dependent Eq. (22) for y require as input the values of θs and λs, the latter equation requires the additional input of the value of E_F [i.e., μ_0, as noted below Eq. (15)] for its solution. We recall that Eq. (16) was derived by assuming that $\mu_0 >> k\,\theta$. Therefore if Eq. (16) is solved for any given values of θs and the λs, and Eq. (22) is solved with the same inputs together with a value of μ_0 which is much greater than k θ, then the solutions of both the equations should yield the same value for y. To carry out this test in detail, we solve Eq. (22) for values of θs as in Eq. (6) and those of λs as in Eq. (26) for the following seven cases: $\lambda_2 = \lambda_3 = 0$, $\lambda_3 = \lambda_1 = 0$, $\lambda_1 = \lambda_2 = 0$, $\lambda_1 = 0$, $\lambda_2 = 0$, $\lambda_3 = 0$, and $\lambda_1 \neq \lambda_2 \neq \lambda_3 \neq 0$. With the same values of the θs and λs and $\mu_0 = 100\,k\,\theta$, solutions of Eq. (22) yield exactly the same results as were obtained via Eq. (16). Some of the values so-obtained are as follows: when $\lambda_2 = \lambda_3 = 0$ (1PEM scenario), y = 9.458; when $\lambda_2 = 0$ (2PEM scenario), y = 1.661; and when none of the λs = 0 (3PEM scenario), y = 1. 195. Having thus shown that Eq. (22) is a reasonable generalization of Eq. (16) for arbitrary values of μ_0, we now solve it for μ_0 (= E_F) as given in Eq. (27) and values of {λ, θ} for each of the seven cases just noted. The results of these calculations are given in **Table 1**.

S. no.	$\lambda_1\lambda_2\lambda_3$	$\theta_1\theta_2\theta_3$ (K)	|W| (meV)	T_c (K)	y	v_0 x10^5 (cms^{-1})	j_0 x10^7 (Acm^{-2})	Remark
1	0.3123 0 0	237 - -	1.70	8.5	9.688	1.41	0.9	1PEM (Ca)
2	0 0.4993 0	- 286 -	7.40	44.4	2.726	4.99	3.2	1PEM (Sr)
3	0 0 0.5000	- - 269	7.01	28.5	2.886	4.72	3.0	1PEM (Bi)
4	0 0.4993 0.5000	- 286 269	26.3	76.8	1.326	10.3	6.6	2PEM (Sr + Bi)
5	0.3123 0 0.5000	237 - 269	17.4	54.1	1.671	8.15	5.2	2PEM (Ca + Bi)
6	0.3123 0.4993 0	237 286 -	18.0	62.4	1.613	8.44	5.4	2PEM (Ca + Sr)
7	0.3123 0.4993 0.5000	237 286 269	38	95	1.199	11.4	7.3	3PEM (Ca + Sr. + Bi)

|W| and T_c are obtained by solving Eqs. (14) and (15), respectively, with the input of $\mu_0 = 1.95\,k\,\theta$ ($\theta = 237$ K), $\mu_1 = 0.105\,\mu_0$, different combinations of λ-values as given in Eq. (26), and the corresponding θ-values given in Eq. (6). For each set of {μ_0, θ, λ}, y is obtained by solving Eq. (22). With $E_F = \mu_0$, $\gamma = (8/15)\times10^{-3}$ J, $v_g = 9.05$ cm^3/g-at, and $s = m^*/m_e = 4.97$, vide Eq. (17), and $n_s = 4.0 \times 10^{20}$/cm^3, vide Eq. (18). v_0 and j_0 corresponding to each value of y are obtained via Eqs. (20) and (21), respectively, with θ = 237 K.

Table 1.
Values of various superconducting parameters of $Bi_2Sr_2CaCu_2O_8$ obtained via GBCSEs.

4.5 The multiple values of the set $\{|W| = \Delta, T_c, j_0\}$ obtained via GBCSEs

Given in **Table 1** are also the values of the set $\{|W| = \Delta, T_c, j_0\}$ and the constituents of j_0 corresponding to each of the seven choices of λs mentioned above and the values of μs given in Eq. (27). Among these, the values of |W| and T_c are obtained via Eqs. (14) and (15), respectively. With E_F (= μ_0) fixed as in Eq. (27) and the values of γ and v_g as noted in the legend for **Table 1**, s and n_s are calculated via Eqs. (17) and (18), respectively. The values of v_0 and j_0 in each row of **Table 1** are obtained via Eqs. (20) and (21), respectively, with the input of θ = 237 K, the value of y noted in it, and the common values of γ, v_g, and E_F.

Excepting |W| = 18 meV in the sixth row, and |W| = 38 meV and T_c = 95 K in the last row of **Table 1**, which were employed as inputs to obtain the three λs, the values of all the other parameters in the Table are predictions following from our approach. Among these, T_c = 62.4 K, corresponding to |W| = 18 meV in the sixth row, is in excellent agreement with the experimental value, whereas the values of the subset {T_c (K), |W| (meV)} = {76.8, 26.3} in the fourth row are in reasonable agreement with the experimental values of {86, 28} [see Eq. (23)]. As concerns j_0, we recall that the experimental value of 3.2 x 10^7 A cm^{-2} noted for it in Eq. (24) corresponds to T = 4.2 K and H = 12 Tesla. Since our value of j_0 = 7.3 x 10^7 A cm^{-2} (in the last row of the Table) corresponds to T = H = 0, it too may be regarded as in reasonably good agreement with the experimental value. Since our approach also provides values of the constituent parameters of j_0, such as n_s and v_0, there is a need to monitor these parameters via experiment for its further validation.

We have so far been concerned with the values of {Tc. Δ} as noted in Eq. (23). The significance of the additional values of T_c that GBCSEs have led to, as seen from **Table 1**, is as follows: these are the temperatures at which CPs bound via one- or more-phonon exchange mechanism breakup. Any such T_c corresponds to vanishing of the associated gap noted in the same row. One would thus expect to see, via appropriately sensitized experimental setups, some of these T_cs in the resistance vs. temperature plot of the SC. We believe that this expectation is met via the experimental results reported in [24] (reproduced in [25]) where, in a plot of resistance vs. temperature for the Bi-Sr-Ca-Cu-O system obtained via a setup sensitized to determine T_cs ≥ 75 K, there occur discontinuities at 75, 80, 85, 108, and 114 K. While these discontinuities are generally attributed to the presence of more than one phase in the SC, they may well alternatively/additionally be due to the breakdown of one or the other mechanism responsible for pairing.

4.6 Are the results given in Table 1 stable?

To address the above question, which is as relevant in the present context as it is for the numerical solution of any problem, we repeated the entire exercise carried out above by taking the Δ- and the T_c-values of the SC as

$$\Delta_2 = 17.5 \text{ meV}, \Delta_3 = 37.5 \text{ meV}, T_{c3} = 94.5 \text{ K}, \quad (28)$$

which deviate slightly from those considered earlier and noted in Eq. (25). A brief account of the results following from Eq. (28) is as follows:

i. In order to satisfy the Bogoliubov constraint, we found that we needed

$$\mu_0 = 2.4\, k\, \theta = 49.02 \text{ meV } (\theta = 237K) \text{ and } \mu_1 = 0.1\, \mu_0,$$

i. which led to

$$\{\lambda_{Ca}, \lambda_{Sr}, \lambda_{Bi}\} = \{0.3628, 0.4338, 0.4918\},$$

i. and to values of y and j0 in the 1-, 2-, and 3-PEM scenarios as

$$\begin{gathered}\lambda_2 = \lambda_3 = 0 \text{ (1PEM)}, y = 6.353, j_0 = 1.58x10^7 \text{ A/cm}^2 \\ \lambda_1 = 0 \text{ (2PEM)}, y = 1.39, j_0 = 7.21x10^7 \text{ A/cm}^2 \\ \lambda_1 \neq \lambda_2 \neq \lambda_3 \neq 0 \text{ (3PEM)}, y = 1.214, j_0 = 8.24x10^7 \text{ A/cm}^2.\end{gathered}$$

Since the sets of values given in Eqs. (25) and (28) lead to values of Δs, T_c, and j_0 in the same ball park, it is seen that our results are stable with respect to small variations in the input variables.

5. Some other applications of GBCSEs

5.1 La_2CuO_4

La_2CuO_4 is unique not only because it heralded the era of high-T_c superconductivity but also because explanation of its $T_c \simeq 38$ K via GBCSEs requires invoking 2PEM, whereas it has only kind of ions (La) that can cause pairing. This paradoxical situation is resolved by appealing to the structure of the unit cell of the SC and employing Eqs. (4) and (5) to determine θ_{La}. Since the unit cell comprises layers of LaO and OLa, if La is considered as the upper bob of the double pendulum in one of these, then it must be the lower bob in the other. Eqs. (4) and (5) therefore lead to two values of θ_{La}. We thus have 2PEM in operation due to $\{\lambda, \theta_{La1}\}$ and $\{\lambda, \theta_{La2}\}$. Following this approach [10], we were able to account for all the values of $(2\Delta_0/K_BT_C)$, i.e., 4.3, 7.1, and 9.3, which were reported by Bednorz and Müller in their Nobel Lecture in 1987.

5.2 Heavy fermion SCs (HFSCs)

Superconductivity in HFSCs—so-named because a conduction electron in them behaves as if it has an effective mass up to three orders of magnitude greater than its free mass—was discovered by Steglich et al. [26]. Even though the highest T_c reported so far for this family of compounds is only 2.3 K (for $CeCoIn_5$), these SCs have been the subject of avid study by theoreticians because of their following unusual properties:

$$E_F < k\theta, kT_c < E_F, T_c/T_F \approx T_F/\theta \approx 0.05, \tag{29}$$

where T_F is the Fermi temperature. Besides the properties noted above, HFSCs are characterized by (a) large heat capacities of conduction electrons, much larger than those found for elemental SCs and about as large as those associated with fixed magnetic momenta, and (b) anisotropy of their gap structures. These features caused the term *exotic* or *unconventional* to be coined for them and led to the revival of an old idea that superconductivity may also arise due to the exchange of magnons, rather than just phonons. As a follow-up of this idea, it was shown in three well-known papers [27–29] that several experimental features of HFSCs can be explained if one assumes that magnetic fluctuations are the cause of d-wave pairing in them. By and large, this seems to be the currently popular view about these SCs.

One of the reasons for regarding HFSCs as outside the purview of BCS theory is the conflict between inequality (8), which is a tenet of the theory, and inequalities (29) which the HFSCs are found to satisfy. Since μ-incorporated GBCSEs are not constrained by inequality (8), we employed them for a detailed study of $CeCoIn_5$, which is a prominent member of the HFSC family. Salient features of this study are as follows. (i) The pairing mechanism was assumed to be 1PEM because we are dealing with a low value of T_c. (ii) Since 1PEM can be due to either Co or Ce ions, we considered both of these possibilities. (iii) Using Eqs. (4) and (5) and θ ($CeCoIn_5$) = 161 K, we found θ_{Ce} = 73 (276) K, corresponding to Ce as the lower (upper) bob in the double pendulum in the layer containing $CeIn_3$, and θ_{Co} = 98.7, (294) K (in the layer containing $CoIn_3$). (iv) Upon solving the μ-incorporated GBCSE for Δ, and different values of μ in an appropriate range determined via a consideration of inequalities (29), we were led to a multitude of Δ values in the range (2.76–3.49) $\times 10^{-4}$ eV corresponding to the single value of T_c = 2.3 K. For the SC under consideration, this is in qualitative accord with the experimental finding via the Bogoliubov quasiparticle interference (QPI) technique [the highest Δ value thus found is $(5.5 \pm 0.05) \times 10^{-4}$ eV, whereas our corresponding value based on the mean-field approximation is 3.49×10^{-4} eV]. (v) Some other results $\left[n_s = (1/3\pi^2)(2m\mu_1/\hbar^2)^{3/2}\right]$ are as follows:

$$7.84x10^{21}\ (\mu_1 = 23.81\ \text{meV}, s = 60.4) \geq n_s\ (\text{cm}^{-3}) \geq 1.07x10^{20}\ (\mu_1 = 0.324\ \text{meV}, s = 253)$$
$$2.59x10^{22}\ (\mu_1 = 25.34\ \text{meV}, s = 126) \geq n_s\ (\text{cm}^{-3}) \geq 3.31x10^{20}\ (\mu_1 = 0.324\ \text{meV}, s = 539),$$

where the upper row corresponds to pairing via the Ce ions and the lower row to pairing via the Co ions.

It follows from even the brief account given above that the framework of μ-incorporated GBCSEs provides a viable explanation of the empirical features of $CeCoIn_5$ as an alternative to the "popular view" about them; for a more detailed treatment of the SC, we refer the reader to [30].

5.3 Fe-based SCs

The salient superconducting properties of $Ba_{0.6}K_{0.4}Fe_2As_2$, which are generic of the Fe-based SCs (also called iron-pnictide SCs), are as follows (for sources of these properties, see [31]): (i) Superconductivity in this SC is due to the $s^{\pm}$-wave state, which signifies that a gap below the Fermi surface and the one corresponding to it above the Fermi surface, Δ_h and Δ_e, respectively, has opposite signs. (ii) In general, $\Delta_h \neq \Delta_e$. (iii) T_c = 37 K. (iv) The highest T_c reported for this class of SCs > 50 K. (v) Prominent Δ-values: 6, 12 meV. (vi) Near-zero values of Δ: Δ for the SC can fall to zero along lines of its Fermi surface. (vii) Some other reported values of Δ (meV): 2.5, 9.0; 3.3, 7.6; 3.6, 8.5, 9.2; 4, 7, 12, 9.5. (viii) j_0: exceeds 0.1 x 10^6 Acm^{-2} at T = 4.2 K and H = 0; 1.1 x 10^7 Acm^{-2} at T = 2 K and H = 0. (ix) A characteristic ratio for the SC: $E_F/kT_c = 4.4$. (x) One of the s = m^*/m_e values: 9.0 (based on ARPES and a four-band model). (xi) Coherence length ξ = 9–14 Å. (xii) The T_c of the SC plotted against a tuning variable has a dome-like structure.

By adopting the framework of GBCSEs as outlined above, it has been shown in [31] that one can quantitatively explain *all* the above features of $Ba_{0.6}K_{0.4}Fe_2As_2$—without invoking a new superconducting state for the SC. This is an important remark because, on the basis of the multiband approach, it has been suggested in a recent review article [32] that superconductivity in Fe-based SCs is manifestation of a *new* state. One of the reasons for this could well be the fact that the BCS equation for Δ is quadratic in Δ and is therefore unaffected when $\Delta \to -\Delta$. On the other hand, GBCSE for W_1 is linear in this variable and has been derived by assuming that W_1 undergoes a change in signature upon crossing the Fermi surface. This is also a feature of GBCSEs for W_2, etc. Since $s^{\pm}$-wave is an inbuilt feature of GBCSEs, one does not have to invent a new state for any SC, as has been suggested in [32].

5.4 Isotope-like effect for composite SCs

The following relation between T_c and the average mass of ions M in an elemental SC is well-known as the isotope effect:

$$T_c \infty M^{-\alpha}. \tag{30}$$

While in the BCS theory $\alpha = 0.5$, values significantly different from it have also been found for some elements such as Mo, Os, and Ru, for which $\alpha = 0.33$, 0.2, and 0, respectively. Therefore, while Eq. (30) does not have the status of a *law*, it nonetheless helped in the formulation of BCS theory because it sheds light on the role of the ion lattice in the scenario of 1PEM, which has been shown to be the operative mechanism for elemental SCs.

We now draw attention to the fact that when Bi and Sr. in $Bi_2Sr_2CaCu_2O_8$ are replaced by Tl and Ba, respectively, the T_c of the SC increases from 95 to 110 K. Consequent upon these substitutions, the only property that can be unequivocally determined is the mass of the SC. Because the operative mechanism for pairing in composite SCs is not 1PEM, it becomes interesting to ask if Eq. (30) can be generalized to address the scenarios of 2PEM and 3PEM. A generalization of Eq. (30) that suggests itself naturally for the 2PEM scenario is

$$T_c = p(M_1M_2)^{-\alpha}, \tag{31}$$

where p is the constant of proportionality and M_1 and M_2 are the masses of ion species that cause pairing. Assuming that the value of $T_c = 95$ K for $Bi_2Sr_2CaCu_2O_8$ is due to 2PEM as in [25], an explanation for the increase in the T_c of $Bi_2Sr_2CaCu_2O_8$ when $Tl_2Ba_2CaCu_2O_8$ is obtained from it via substitutions was given in [33]. Noting that Bi and Tl belong to the same period and Sr. and Ba to the same group of the periodic table, several suggestions were made in [33] to further increase the T_c of the Bi-based SC. An example, following from our study, it was shown that T_c (Bi_2 Mg_2 $CaCu_2O_8$) should be 171 K.

6. Discussion

The T_cs and Δs of most of the hetero-structured, multi-gapped SCs which were studied above via the GBCSE-based approach have also been studied via the more-widely followed multiband approach (MBA) which originated with the work of Suhl et al. [34]. Because the former approach sheds light on additional features of these SCs, such as j_0, s, n_s, etc., as also because of its distinctly different conceptual basis, it complements the latter approach. Since the conceptual bases of both the approaches have been dealt with in detail in a recent paper [35], we confine ourselves here to the following remarks:

i. Employing the concept of a superpropagator, the GBCSE-based approach invariably invokes a λ for each of the ion species in an SC that may cause pairing. One then has the same λs in the equations for any Δ and the corresponding T_c of the SC—as is the case for elemental SCs. Multiple gaps arise in this approach because different combinations of λs operate on different parts of the Fermi surface due to its undulations. For a discussion of the moot question concerning the assumption of *locally* spherical values characterizing the Fermi surface of a composite SC, see [33].

On the other hand, the number of bands employed in MBA for the same SC differs from author to author. In this approach, in the two-band models, multiple gaps arise because the Hamiltonian is now postulated to have a term for pairing in each of the bands and another term corresponding to crossband pairing. For the T_c of the SC, one often employs the Migdal-Eliashberg-McMillan approach [36] which, even though based on 1PEM, allows λ to be greater than unity because it is based on an integral equation and the expansion parameter of which is not λ but m_e/M, where M is the mass of an ion species.

ii. Because of the feature of the GBCSE-based approach discussed above, with the input of the values of any two gaps of an SC and a value of T_c, it goes on to shed light on several other values of these parameters. This is not so for MBA.

7. Conclusions

i. Besides the SCs dealt with above, the GBCSE-based approach has been applied for the study of several elemental SCs, MgB_2, YBCO and Tl-2212 [10, 19, 20], $SrTiO_3$ [10], and NbN [21]. It is hence seen that a striking feature of this approach is its versatility: it is applicable to a wide variety of SCs that includes the so-called exotic or unconventional SCs such as the HFSCs and the Fe-based SCs, without invoking a new superconducting state for the latter.

ii. We believe to have shown that in explaining the empirical features of high-T_c SCs, the GBCSE-based approach goes farther than MBA or, so far as we are aware, any other approach, e.g., [37].

iii. The fact that T_c ($Bi_2Sr_2Ca_2Cu_3O_{10}$) > T_c ($Bi_2Sr_2CaCu_2O_8$) > T_c ($YBa_2Cu_3O_7$) > T_c (MgB_2), etc., obviously suggests that the greater the complexity of structure of the SC, the greater is its T_c owing to the increase in the number of channels that may cause formation of CPs. While this is a situation that the GBCSE-based approach can easily deal with, in practice, going beyond $Bi_2Sr_2Ca_2Cu_3O_{10}$, for example, is most likely to lead to an unstable and hence unrealizable SC. This is a problem that we believe belongs to the realm of chemical engineering.

iv. It was noted above that 14 different values of j_0 have been reported for Bi-2212. For none of these values were reported the values of the other superconducting parameters of the SC. On the basis of [19, 20], we believe that, in order to help theory to suggest means to increase the T_c of an SC, the report of any of its properties, for example, j_0, should be accompanied by, in so far as it is experimentally feasible, a list of all its other superconducting properties, viz., θ, T_c, Δs, m^*, v_0, n_s, γ, and v_g.

v. The E_F-incorporated GBCSEs in this communication correspond to the T_c of a composite SC and its Δ-values at T = H = 0. These equations can be further generalized to deal with the situation when the SC is in heat bath in an external magnetic field, i.e., when $T \neq H \neq 0$, a procedure for which has been given in [38]. The import of this remark is that not only will such an undertaking enable theory to address the j_0 values of an SC which are generally reported for such values of T and H but also that it may shed new light on the Volovik effect for the Fe-based SCs, as discussed in, e.g., [32].

Conflict of interest

The authors declare that there is no conflict of interest with regard to this submission.

Author details

Gulshan Prakash Malik[1*†] and Vijaya Shankar Varma[2‡]

1 Sushant Lok 1, Gurgaon, Haryana, India

2 Mall Apartments, Mall Road, Delhi, India

*Address all correspondence to: gulshanpmalik@yahoo.com

†Formerly: Theory Group, School of Environmental Sciences, Jawaharlal Nehru University, New Delhi, India

‡Formerly: Department of Physics and Astrophysics, University of Delhi, Delhi, India

References

[1] Onnes Kamerlingh H. Further experiments with liquid helium. G. On the electrical resistance of pure metals, etc. VI. On the sudden change in the rate at which the resistance of mercury disappears. Communications from the Laboratory of Physics at the University of Leiden. 1911;**124c**:267-271

[2] Bardeen J, Cooper LN, Schrieffer JR. Theory of superconductivity. Physics Review. 1957;**108**:1175-1204. DOI: 10.1103/PhysRev.108.1175

[3] Salpeter EE, Bethe H. A relativistic equation for bound-state problems. Physics Review. 1951;**84**:1232-1242. DOI: 10.1103/PhysRev.84.1232

[4] Matsubara T. A new approach to quantum-statistical mechanics. Progress of Theoretical Physics. 1955;**14**:351-378. DOI: 10.1143/PTP.14.351

[5] Okubo S. Note on the second kind interaction. Progress of Theoretical Physics. 1954;**11**:80-94. DOI: 10.1143/PTP.11.80

[6] Dolan L, Jackiw R. Symmetry behavior at finite temperature. Physics Review. 1974;**D9**:3320-3341. DOI: 10.1103/PhysRevD.9.3320

[7] Biswas SN, Malik GP, Sudarshan ECG. Superpropagator for a nonpolynomial field. Physics Review. 1973;**D7**:2884-2886. DOI: 10.1103/PhysRevD.7.2884

[8] Seitz F. Modern Theory of Solids. NY: McGraw Hill; 1940. 698 p. ISBN-10: 9780070560307; ISBN-13: 978–0070560307; ASIN: 0070560307

[9] Blatt JM. Theory of Superconductivity. NY: Academic Press; 1964. 486 p. QC612.S8B57

[10] Malik GP. Superconductivity – A New Approach Based on the Bethe-Salpeter Equation in the Mean-Field Approximation. Volume 21 in the series on condensed matter physics. Singapore: World Scientific; 2016. 223 p. ISBN 9814733075 (hard cover)

[11] Salpeter EE. Mass corrections to the fine structure of hydrogen-like atoms. Physics Review. 1952;**87**:328-343. DOI: 10.1103/PhysRev.87.328

[12] Lee DH. Iron-based superconductors: Nodal rings. Nature Physics. 2012;**8**: 364-365. DOI: 10.1038/nphys2301

[13] Zhang Y et al. Nodal superconducting-gap structure in ferropnictide superconductor $BaFe_2(As_{0.7}P_{0.3})_2$. Nature Physics. 2012; **8**:371-375. DOI: 10.1038/nphys2248

[14] Allan MP et al. Anisotropic energy gaps of iron-based superconductivity from intra-band Quasiparticle interference in LiFeAs. Science. 2012; **336**:231-236. DOI: 10.1126/science

[15] Alexandrov AS. Nonadiabatic polaronic superconductivity in MgB_2 and cuprates. Physica C: Superconductivity. 2001;**363**:231-236. DOI: 10.1016/S0921-4534(01)01095-4

[16] Jarlborg T, Bianconi A. Fermi surface reconstruction of superoxygenated La2CuO4 superconductors with ordered oxygen interstitials. Physical Review B. 2013;**87**: 054514-054516. DOI: 10.1103/PhysRevB.87.054514

[17] van der Marel D, van Mechelen JLM, Mazin II. Common Fermi-liquid origin of T^2 resistivity and superconductivity in n-type $SrTiO_3$. Physical Review B. 2011;**84**:205111-205111. DOI: 10.1103/PhysRevB.84.205111

[18] Eagles DM. Possible pairing without superconductivity at low carrier

concentrations in bulk and thin-film superconducting semiconductors. Physics Review. 1969;**186**:456-463. DOI: 10.1103/PhysRev.186.456

[19] Malik GP. On the role of fermi energy in determining properties of superconductors: A detailed study of two elemental superconductors (Sn and Pb), a non-cuprate (MgB_2) and three cuprates (YBCO, Bi-2212 and Tl-2212). Journal of Superconductivity and Novel Magnetism. 2016;**29**:2755-2764. DOI: 10.1007/s10948-016-3637-5

[20] Malik GP. Correction to: On the role of fermi energy in determining properties of superconductors: A detailed study of two elemental superconductors (Sn and Pb), a non-cuprate (MgB_2) and three cuprates (YBCO, Bi-2212 and Tl-2212). Journal of Superconductivity and Novel Magnetism. 2018;**31**:941. DOI: 10.1007/s10948-017-4520-8

[21] Malik GP. A detailed study of the role of Fermi energy in determining properties of superconducting NbN. Journal of Modern Physics. 2017;**8**: 99-109. DOI: 10.4236/jmp.2017.81009

[22] Poole CP. Handbook of Superconductivity. San Diego: Academic Press; 2000. 692 p. ISBN: 0–12–561460-8, Library of Congress Card No.: 99–60091

[23] Kametani F, et al. Comparison of growth texture in round Bi2212 and flat Bi2223 wires and its relation to high critical current density development. Scientific Reports. 2015;**5**. Article Number 8285. DOI: 10.1038/srep08285

[24] Hiroshi M, Kazumasa T, editors. Bismuth-Based High-Temperature Superconductors. NY: Marcel-Dekker; 1996. 626 p. ISBN: 0–8247-9690-X (alk. paper)

[25] Malik GP, Malik U. A study of thallium- and bismuth-based high-temperature superconductors in the framework of the generalized BCS equations. Journal of Superconductivity and Novel Magnetism. 2011;**24**:255-260. DOI: 10.1007/s10948-010-1009-0

[26] Steglich F et al. Superconductivity in the presence of strong Pauli Paramagnetism: $CeCu_2Si_2$. Physical Review Letters. 1976;**43**:1892-1896. DOI: 0.1103/PhysRevLett.43.1892

[27] Miyaki K, Rink SS, Varma CM. Spin-fluctuation-mediated even-parity pairing in heavy-fermion superconductors. Physical Review B. 1986;**34**:6554-6556. DOI: 10.1103/PhysRevB.34.6554

[28] Monod MT, Bourbonnias C, Emery V. Possible superconductivity in nearly antiferromagnetic itinerant fermion systems. Physical Review B. 1986;**34**: 7716-7720. DOI: 10.1103/PhysRevB.34.7716

[29] Scalapino DJ, Loh E, Hirsch JE. D-wave pairing near a spin-density-wave instability. Physical Review B. 1986;**34**:8190-8192. DOI: 10.1103/PhysRevB.34.8190

[30] Malik GP. A study of heavy-fermion superconductors via BCS equations incorporating chemical potential. Journal of Modern Physics. 2015;**6**: 1233-1242. DOI: 10.4236/jmp.2015.69128

[31] Malik GP. On the $s^{\pm}$-wave superconductivity in the iron-based superconductors: A perspective based on a detailed study of $Ba_{0.6}K_{0.4}Fe_2As_2$ via the generalized Bardeen-Cooper-Schrieffer equations incorporating Fermi energy. Open Journal of Composite Materials. 2017;**7**:130-145. DOI: 10.4236/ojcm.2017.73008

[32] Bang Y, Stewart GR. Superconducting properties of the $s^{\pm}$-wave state: Fe-based superconductors. Journal of Physics. Condensed Matter.

2017;**29**:123003-123046. DOI: 10.1088/1361-648X/aa564b

[33] Malik GP. On the isotope-like effect for high-T_c superconductors in the scenario of 2-phonon exchange mechanism for pairing. World Journal of Condensed Matter Physics. 2018;**8**: 109-114. DOI: 10.4236/wjcmp.2018.83007

[34] Suhl H, Matthias BT, Walker LR. Bardeen-Cooper-Schrieffer theory in the case of overlapping bands. Physical Review Letters. 1959;**3**:552-554. DOI: 10.1103/PhysRevLett. 3.552

[35] Malik GP. An overview of the multi-band and the generalized BCS equations-based approaches to Deal with hetero-structured superconductors. Open Journal of Microphysics. 2018;**8**:7-13. DOI: 10.4236/ojm.2018.82002

[36] McMillan WL. Transition Temperature of Strong-Coupled Superconductors. 1968;**167**:331-344. DOI: 10.1103/PhysRev.167.331

[37] Freeman J, Sandro Massida JY. Local density theory of high T_c superconductors: $Ba_2Sr_2CaCu_2O_8$. Physica C: Superconductivity. 1988; **153-155**:1225-1226. DOI: 10.1016/0921-4534(88)90253-5

[38] Malik GP. On Landau quantization of Cooper pairs in a heat bath. Physica B: Condensed Matter. 2010;**405**:3475-3481. DOI: 10.1016/j.physb.2010.05.026

Chapter 5

Nematicity in Electron-Doped Iron-Pnictide Superconductors

Hong-Yi Chen

Abstract

The nature of the nematicity in iron pnictides is studied with a proposed magnetic fluctuation. The spin-driven order in the iron-based superconductor has been realized in two categories: stripe SDW state and nematic state. The stripe SDW order opens a gap in the band structure and causes a deformed Fermi surface. The nematic order does not make any gap in the band structure and still deforms the Fermi surface. The electronic mechanism of nematicity is discussed in an effective model by solving the self-consistent Bogoliubov-de Gennes equations. The nematic order can be visualized as crisscross horizontal and vertical stripes. Both stripes have the same period with different magnitudes. The appearance of the orthorhombic magnetic fluctuations generates two uneven pairs of peaks at $(\pm\pi, 0)$ and $(0, \pm\pi)$ in its Fourier transformation. In addition, the nematic order breaks the degeneracy of d_{xz} and d_{yz} orbitals and causes the elliptic Fermi surface near the Γ point. The spatial image of the local density of states reveals a $d_{x2\text{-}y2}$-symmetry form factor density wave.

Keywords: magnetic fluctuation, stripe SDW, nematic order, two-orbital, elliptic Fermi surface, LDOS maps

1. Introduction

The discovery of Fe-based superconductors with critical temperatures up to 55 K has begun a new era of investigations of the unconventional superconductivity. In common with copper-like superconductors (cuprate), the emergency of superconductivity in electron-doped Fe-pnictides such as $Ba(Fe_{1-x}Co_xAs)_2$ is to suppress the magnetic order and fluctuations originated in the parent compound with $x = 0$ [1]. The intertwined phases between the superconductivity and stripe spin density wave (SDW) order (ferromagnetic stripes along one Fe-Fe bond direction that is antiferromagnetically aligned along orthogonal Fe—Fe bond) are of particular interests. In both pnictides and cuprates, the experimentally observed nematicity exists in an exotic phase between the superconductivity (SC) and the stripe SDW [2]. The nematicity occurs in weakly doped iron pnictides with tetragonal-to-orthorhombic structural transition [3–17], i.e., in a square unit cell, the point-group symmetry is reduced from C_4 (tetragonal) to C_2 (orthorhombic).

At present, there are two scenarios for the development of nematic order through the electronic configurations [18]. One scenario is the orbital fluctuations [19–23]. The structural order is driven by orbital ordering. The orbital ordering induces magnetic anisotropy and triggers the magnetic transition at a lower temperature. The other scenario is the spin fluctuation [24–27]. The magnetic mechanism for the structural order is associated with the onset of SDW.

Recently, Lu et al. [28] reported that the low-energy spin fluctuation excitations in underdoped sample $BaFe_{1.915}Ni_{0.085}As_2$ change from C_4 symmetry to C_2 symmetry in the nematic state. Zhang et al. [29] exhibited that the reduction of the spin-spin correlation length at $(0, \pi)$ in $BaFe_{1.935}Ni_{0.065}As_2$ happens just below T_s, suggesting the strong effect of nematic order on low-energy spin fluctuations. Apparently, these experiments above have provided a spin-driven nematicity picture.

The partial melting of SDW has been proposed as the mechanism to explain the nematicity. The properties of the spin-driven nematic order have been studied in Landau-Ginzburg-Wilson's theory [18, 24–26]. Meanwhile, the lack of the realistic microscopic model is responsible for the debates where the leading electronic instability, i.e., the onset of SDW, causes the nematic order. Recently, an extended random phase approximation (RPA) approach in a five-orbital Hubbard model including Hund's rule interaction has shown that the leading instability is the SDW-driven nematic phase [30]. Although the establishment of the nematicity in the normal state has attracted a lot of attentions, the microscopic description of the nematic order and, particularly, the relation between SC and the nematic order are still missing.

The magnetic mechanism for the structural order is usually referred to the Ising-nematic phase where stripe SDW order can be along the x-axis or the y-axis. The nematic phase is characterized by an underlying electronic order that the Z_2 symmetry between the x- and y- directions is broken above and the O(3) spin-rotational symmetry is preserved [25].

The magnetic configuration in FeSCs can be described in terms of two magnetic order parameters Δ_x and Δ_y. Both order parameters conventionally defined in momentum space are written as

$$\Delta_\ell = \sum_k c^\dagger_{k+Q_\ell,\alpha} \sigma_{\alpha\beta} c_{k,\beta}, \tag{1}$$

where $\ell = x$ or y. Here the wave vectors $Q_x = (\pi, 0)$ and $Q_y = (0, \pi)$ correspond to the spins parallel along the y-axis and antiparallel along the x-axis and the spins parallel along the x-axis and antiparallel along the y-axis, respectively.

In the stripe SDW state, the order parameters are set to $\langle \Delta_x \rangle \neq 0$ or $\langle \Delta_y \rangle \neq 0$, i.e., $\langle \Delta_\ell \rangle \neq 0$. This implies to choose an ordering vector either Q_x or Q_y. The Z_2 symmetry indicating to the degenerate of spin stripes along the y-axis (corresponding to Q_x) or x-axis (corresponding to Q_y) is broken. In addition, the O (3) spin-rotational symmetry is also broken. In the real space configuration, the magnetic ground state is an orthorhombic uniaxial stripe state. The stripe order reduces the point-group symmetry of a unit cell from C_4 (tetragonal) to C_2 (orthorhombic). In the nematic state, the order parameters are set to $\langle \Delta_x \rangle = \langle \Delta_y \rangle = 0$ and $\langle \Delta_x^2 \rangle \neq \langle \Delta_y^2 \rangle$. This implies that the magnetic fluctuations associated with one of the ordering vectors are stronger than the other $\langle \Delta_x^2 \rangle > \langle \Delta_y^2 \rangle$ or $\langle \Delta_y^2 \rangle > \langle \Delta_x^2 \rangle$. Therefore, the Z_2 symmetry is broken, but the O(3) spin-rotational symmetry is not. In the real space configuration, the x- and y-directions of the magnetic fluctuations are inequivalent.

Recently, the reentrant C_4 symmetry magnetic orders have been reported in hole-doped Fe-pnictide [27, 31, 32]. A double-Q order (choose both Q_x and Q_y) has been proposed to change the ground state from striped to tetragonal [33, 34].

Two stripe orders with the ordering vectors Q_x and Q_y are superposed to preserve the tetragonal symmetry. As a matter of fact, the nematic phase is characterized by an underlying electronic order that spontaneously breaks tetragonal symmetry. Since the double-Q order does not break the C_4 symmetry, it is not suitable to explain the nematicity.

The magnetic fluctuations trigger a transition from the tetragonal-to-orthorhombic phase. At very high temperature, $T>T_S$, $\langle \Delta_x \rangle = \langle \Delta_y \rangle = 0$, and the fluctuations of all order parameters have equal strength, i.e., $\langle \Delta_x^2 \rangle = \langle \Delta_y^2 \rangle$. As the temperature lowers, $T_N < T < T_S$, the thermodynamic average of order parameters still remains $\langle \Delta_x \rangle = \langle \Delta_y \rangle = 0$ but $\langle \Delta_x^2 \rangle \neq \langle \Delta_y^2 \rangle$. The fluctuations of one of the orders Δ_x are on average different from the fluctuations of the other order Δ_y implying a broken Z_2 symmetry and a preserved $O(3)$ symmetry. When the temperature is below T_N, the magnetic ground state is a stripe SDW state, i.e., $\langle \Delta_x \rangle = 0$ or $\langle \Delta_y \rangle = 0$.

In this chapter, we will exploit a two-orbital model to study the interplay between SC and nematicity in a two-dimensional lattice. The two-orbital model has been successfully used in many studies such as quasiparticle excitation, the density of states near an impurity [35, 36] and the magnetic structure of a vortex core [37].

2. Model

Superconductivity in the iron-pnictide superconductors originates from the FeAs layer. The Fe atoms form a square lattice, and the As atoms are alternatively above and below the Fe-Fe plane. This leads to two sublattices of irons denoted by sublattices A and B. Many tight-binding Hamiltonians have been proposed to study the electronic band structure that includes five Fe 3d orbitals [38], three Fe orbitals [39, 40], and simply two Fe bands [41–43]. Each of these models has its own advantages and range of convenience for calculations. For example, the five-orbital tight-binding model can capture all details of the DFT bands across the Fermi energy in the first Brillouin zone. However, in practice, it becomes a formidable task to solve the Hamiltonian with a large size of lattice in real space even in the mean-field level. Several studies used five-orbital models in momentum space to investigate the single-impurity problem for different iron-based compounds such as $LaFeAsO_{1-x}F_x$, LiFeAs, and $K_xFe_{2-y}Se_2$. These studies confirmed that the detail of electronic bands strongly influences on the magnitude and location of the in-gap resonant states generated by the scattering of quasiparticles from single impurity [44–46].

On the other hand, the two-orbital models apparently have a numerical advantage dealing with a large size of lattice while retaining some of the orbital characters of the low-energy bands. Among the two-orbital (d_{xz} and d_{yz}) models, Zhang [47] proposed a phenomenological approach to take into account the two Fe atoms per unit cell and the asymmetry of the As atoms below and above of the Fe plane. Later on, Tai and co-workers [48] improved Zhang's model by a phenomenological two-by-two-orbital model (two Fe sites with two orbitals each). The obtained low-energy electronic dispersion agrees qualitatively well with DFT in LDA calculations of the entire Brillouin zone of the 122 compounds.

The multi-orbital Hamiltonian of the iron-pnictide superconductors in a two-dimensional square lattice is described as

$$H = \sum_{ijuv\alpha} t_{ijuv} c^{\dagger}_{iu\alpha} c_{jv\alpha} - \mu \sum_{iu\alpha} n_{iu\alpha} + U \sum_{iu} n_{iu\uparrow} n_{iu\downarrow} + \left(U' - \frac{J_H}{2}\right) \sum_{i,\, u<v,\, \alpha\beta} n_{iu\alpha} n_{iv\beta} \\ - 2J_H \sum_{i,\, u<v} S_{iu} \cdot S_{iv} + J' \sum_{i,\, u\neq v} c^{\dagger}_{iu\uparrow} c^{\dagger}_{iu\downarrow} c_{iv\downarrow} c_{iv\uparrow}, \tag{2}$$

where

$$n_{iu\alpha} = c^{\dagger}_{iu\alpha} c_{iu\alpha}, \tag{3}$$

$$S_{iu} = \frac{1}{2} \sum_{\alpha\beta} c^{\dagger}_{iu\alpha} \sigma_{\alpha\beta} c_{iu\beta}, \tag{4}$$

with $\sigma_{\alpha\beta}$ the Pauli matrices. The operators $c^{\dagger}_{iu\alpha}$ ($c_{iu\alpha}$) create (annihilate) an ectron with spin $\alpha, \beta = \uparrow, \downarrow$ in the orbital $u, v = 1, 2$ on the lattice site i; t_{ijuv} is the hopping matrix element between the neighbor sites, and μ is the chemical potential. U (U') is the intraorbital (interorbital) on-site interaction. Hund's rule coupling is J_H and the pair hopping energy is J'. The spin-rotation invariance gives $U' = U - 2J_H$ and $J' = J_H$ [49]. Repulsion between electrons requires $J_H < U/3$.

Here, we adopt Tai's phenomenological two-by-two-orbital model because it is able to deal with a large size of lattice in many aspects and the details of low-energy bands are similar to the results from DFT + LDA. In a two-orbital model, the hopping amplitudes are chosen as shown in **Figure 1** [48] to fit the band structure from the first-principle calculations:

$$t_1 = t_{i,\, i\pm x(y),\, u,\, u} = -1, \tag{5}$$

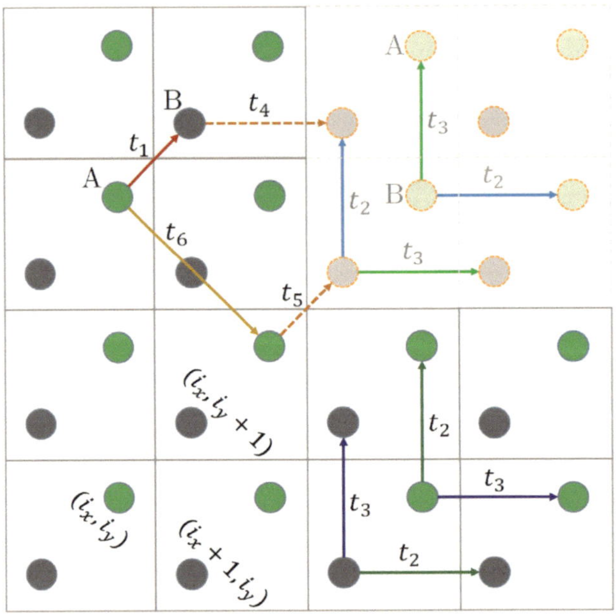

Figure 1.
(color online) two-dimensional square lattice of the iron-based superconductors. There are two Fe atoms (green and gray color) in a unit cell, and each atom has two orbitals. The bright color circle represents the first orbital, and the faded color circle represents the second orbital. Solid lines indicate the hopping between the atoms in the same orbital and dashed lines indicate the hopping between the atoms in the different orbitals.

$$\begin{cases} \frac{1+(-1)^{x+y+u}}{2} t_2 + \frac{1-(-1)^{x+y+u}}{2} t_3 = t_{i,\, i\pm(x+y),\, u,\, u} = 0.08 \\ \frac{1+(-1)^{x-y+u}}{2} t_3 + \frac{1-(-1)^{x-y+u}}{2} t_2 = t_{i,\, i\pm(x-y),\, u,\, u} = 1.35 \end{cases}, \tag{6}$$

$$t_4 = t_{i,\, i\pm(x\pm y),\, u,\, v\neq u} = -0.12, \tag{7}$$

$$t_5 = t_{i,\, i\pm x(y),\, u,\, v\neq u} = 0.09, \tag{8}$$

$$t_6 = t_{i,\, i\pm 2x(2y),\, u,\, u} = 0.25, \tag{9}$$

where $u \neq v$ indicates two different orbitals.

Figure 1 shows the hopping parameters between unit cells and orbitals. For the same orbital, the hopping parameters t_2 and t_3 are chosen differently along the mutually perpendicular directions. The C_4 symmetry on the same orbital between different sublattices is broken. However, t_2 and t_3 are twisted for the different Fe atoms on the same sublattice which restore the C_4 symmetry of the lattice structure.

In the mean-field level

$$\mathrm{H} = \mathrm{H}_0 + \mathrm{H}_\Delta + \mathrm{H}_{\mathrm{int}}, \tag{10}$$

the Hamiltonian is self-consistently solved accompanied with s^{+-}-wave superconducting order. The mean-field scheme is the same as Ref. [48]:

$$\mathrm{H}_0 = \sum_{ijuv\alpha} t_{ijuv} c^\dagger_{iu\alpha} c_{jv\alpha} - \mu \sum_{iu\alpha} n_{iu\alpha}, \tag{11}$$

$$\mathrm{H}_\Delta = \sum_{iju\alpha\beta} \left(\Delta_{iju} c^\dagger_{iu\alpha} c^\dagger_{ju\beta} + \mathrm{h.c.} \right), \tag{12}$$

$$\mathrm{H}_{\mathrm{int}} = U \sum_{iu,\, \alpha\neq\beta} \langle n_{iu\beta} \rangle n_{iu\alpha} + U' \sum_{i,\, u<v,\, \alpha\neq\beta} \langle n_{iu\beta} \rangle n_{iv\alpha} + (U' - J_H) \sum_{i,\, u<v,\, \alpha} \langle n_{iu\alpha} \rangle n_{iv\alpha}. \tag{13}$$

The next nearest-neighbor intraorbital attractive interaction V is responsible for the superconducting order parameter $\Delta_{ijuu} = V\langle c_{i\uparrow} c_{j\downarrow} \rangle$ [50–52]. According to the literatures where U is chosen to be 3.2 or less [48, 52], the magnetic configuration is a uniform SDW order. In order to make the stripe SDW order and the nematic order by changing electron doping, i.e., the chemical potential μ, we choose the parameters of interactions $U = 3.5, J_H = 0.4$, and $V = 1.3$ to induce a nematic order within a small doping range.

In momentum space, the spin configuration is determined by the order parameters Δ_x and Δ_y. The combination of these order parameters lacks the visualization of the magnetic structure in real space. Therefore, we choose the staggered magnetization M_i in a lattice to study the states driven by the magnetic mechanism. The magnetic configuration is described as

$$M_i = M_1 \cos\left(q_y \cdot r_i\right) e^{iQ_x \cdot r_i} + M_2 \sin\left(q_x \cdot r_i\right) e^{iQ_y \cdot r_i}, \tag{14}$$

where the wave vectors $q_x = (2\pi/\lambda, 0)$ and $q_y = (0, 2\pi/\lambda)$ correspond to a modulation along the x-axis and the y-axis with wavelength λ. M_1 and M_2 are the amplitude of the modulation.

In the case of the absence of both q_x and q_y, M_i becomes

$$M_i = M_1 e^{iQ_x \cdot r_i} + M_2 e^{iQ_y \cdot r_i}. \tag{15}$$

As $M_1 = 0$ or $M_2 = 0$ is chosen, i.e., choosing the ordering vector either Q_x or Q_y, the state is a stripe SDW state. The existence of q_x and q_y does not affect the stripe SDW state. As two values of M_1 and M_2 are arbitrarily chosen, the spin configuration forms a stripe SDW along the $(1, 1)$ direction. The existence of q_x and q_y has a lot of influences on the nematic state.

In the nematic state, the presence of both Q and q ordering vectors is necessary. Unlike the double-Q model, with the choice $M_1 \neq M_2$, the magnitudes of the modulated antiparallel spins along the x-axis and the y-axis are different due to the existence of q vectors. The magnetic configuration is attributed to two inequivalent stripes interpenetrating each other and formed a checked pattern. The checked pattern has the same period along the x- and y-directions. The period is determined by the value of q in the periodic boundary conditions. The value of q, therefore, cannot be arbitrary and must commensurate the lattice to stabilize the modulation and lower the energy of the system. As two modulated stripes have no phase difference, the checked patter shows a s-wave-like symmetry. In addition, as the modulations have a phase shift of $\pi/2$, the checked pattern shows a d-wave-like symmetry.

Figure 2 displays the Fermi surface and the band structure in the absence of SDW at the normal state, i.e., the superconductivity is set to zero. In the absence of SDW $(U = 0)$, the Hamiltonian in momentum space is

$$\Psi^\dagger = \left(c^\dagger_{A1k\sigma}\; c^\dagger_{A2k\sigma}\; c^\dagger_{B1k\sigma}\; c^\dagger_{B2k\sigma}\right), \tag{16}$$

$$\mathrm{H}_0 = \sum_{k\sigma} \Psi^\dagger \begin{pmatrix} \varepsilon_{A1} & \varepsilon_{12} & \varepsilon_{AB} & \varepsilon_c \\ \varepsilon_{12} & \varepsilon_{A2} & \varepsilon_c & \varepsilon_{AB} \\ \varepsilon_{AB} & \varepsilon_c & \varepsilon_{B1} & \varepsilon_{12} \\ \varepsilon_c & \varepsilon_{AB} & \varepsilon_{12} & \varepsilon_{B2} \end{pmatrix} \Psi, \tag{17}$$

where

$$\varepsilon_{A1} = -2t_3 \cos k_x - 2t_2 \cos k_y - 4t_6 \cos k_x \cos k_y, \tag{18}$$

$$\varepsilon_{A2} = -2t_2 \cos k_x - 2t_3 \cos k_y - 4t_6 \cos k_x \cos k_y, \tag{19}$$

$$\varepsilon_{B1} = -2t_2 \cos k_x - 2t_3 \cos k_y - 4t_6 \cos k_x \cos k_y, \tag{20}$$

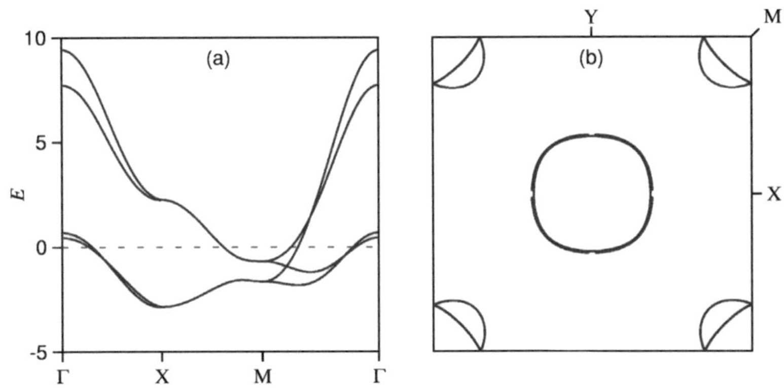

Figure 2.
(color online) (a) and (b) are, respectively, the band structure and the Fermi surface without SDW. The Fermi energy (red dashed line) corresponds to the electron filling n = 2.1.

$$\varepsilon_{B2} = -2t_3 \cos k_x - 2t_2 \cos k_y - 4t_6 \cos k_x \cos k_y, \tag{21}$$

$$\varepsilon_{12} = t - 2_4 \cos k_x - 2t_4 \cos k_y, \tag{22}$$

$$\varepsilon_{AB} = -4t_1 \cos \frac{k_x}{2} \cos \frac{k_y}{2}, \tag{23}$$

$$\varepsilon_c = -4t_5 \cos \frac{k_x}{2} \cos \frac{k_y}{2}. \tag{24}$$

The eigenvalues are

$$E_{1,k} = \overline{\varepsilon}_+ - \sqrt{\overline{\varepsilon}_-^2 + (\varepsilon_{12} - \varepsilon_{AB})^2} - \varepsilon_c - \mu, \tag{25}$$

$$E_{2,k} = \overline{\varepsilon}_+ + \sqrt{\overline{\varepsilon}_-^2 + (\varepsilon_{12} - \varepsilon_{AB})^2} - \varepsilon_c - \mu, \tag{26}$$

$$E_{3,k} = \overline{\varepsilon}_+ - \sqrt{\overline{\varepsilon}_-^2 + (\varepsilon_{12} + \varepsilon_{AB})^2} + \varepsilon_c - \mu, \tag{27}$$

$$E_{4,k} = \overline{\varepsilon}_+ + \sqrt{\overline{\varepsilon}_-^2 + (\varepsilon_{12} + \varepsilon_{AB})^2} + \varepsilon_c - \mu, \tag{28}$$

where

$$\overline{\varepsilon}_\pm = \frac{\varepsilon_{A1} \pm \varepsilon_{B1}}{2}. \tag{29}$$

Figure 2(a) shows that two hole bands are around the Γ point and two electron bands are around the M point. There is no gap in the band structure where the superconductivity is able to open a gap. The nature of Fermi surface is revealed by the line at the Fermi energy crossing the band dispersion through $\Gamma - X - M - \Gamma$ points. **Figure 2(b)** displays that the Fermi surface contains two hole pockets that are centered at Γ point and two electron pockets that are centered at M points. In addition, the Fermi surface also exhibits the C_4 symmetry of the lattice structure.

In the stripe SDW state, the spin configuration is shown as **Figure 3**. The stripe SDW order enlarges the two-Fe unit cell to four-Fe unit cell as denoted by the blue

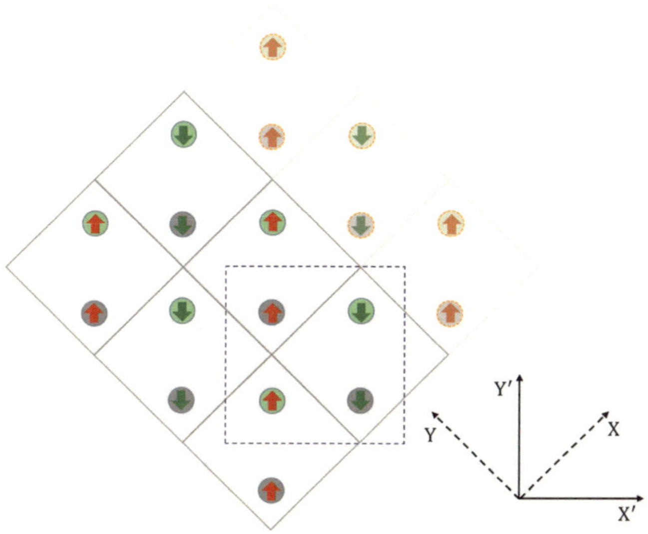

Figure 3.
(color online) the schematic lattice structure of the Fe layer in the stripe SDW state. Blue dashed square denotes the four-Fe unit cell in the stripe SDW state.

dashed square in **Figure 3**. The antiferromagnetic order is along the X′-axis and the ferromagnetic order is along the Y′-axis. The magnetic unit cell (four-Fe unit cell) with lattice spacing $\sqrt{2}a$ is oriented at a 45° angle with respect to the nonmagnetic unit cell (two-Fe unit cell). The magnetic Brillouin zone is a square oriented at a 45° angle with respect to the crystal Brillouin zone. The size of the magnetic unit cell is twice of the nonmagnetic unit cell, and the size of the magnetic Brillouin zone is half of the crystal Brillouin zone. The Hamiltonian in momentum space is

$$\widetilde{\Psi}_k^\dagger = \left(c^\dagger_{A_1^{(1)}k} \quad c^\dagger_{A_1^{(2)}k} \quad c^\dagger_{B_1^{(1)}k} \quad c^\dagger_{B_1^{(2)}k} \quad c^\dagger_{B_2^{(1)}k} \quad c^\dagger_{B_2^{(2)}k} \quad c^\dagger_{A_2^{(1)}k} \quad c^\dagger_{A_2^{(2)}k}\right), \tag{30}$$

$$\mathrm{H} = \sum_k \widetilde{\Psi}_k^\dagger \cdot \mathbf{h} \cdot \widetilde{\Psi}_k, \tag{31}$$

$$\mathbf{h} = \begin{pmatrix} \varepsilon_{A_1^1} & 0 & \varepsilon_{t_1^x} & \varepsilon_{t_5^x} & \varepsilon_{t_1^y} & \varepsilon_{t_5^y} & \varepsilon_{t23} & \varepsilon_{t4} \\ 0 & \varepsilon_{A_1^2} & \varepsilon_{t_5^x} & \varepsilon_{t_1^x} & \varepsilon_{t_5^y} & \varepsilon_{t_1^y} & \varepsilon_{t4} & \varepsilon_{t32} \\ \varepsilon_{t_1^x} & \varepsilon_{t_5^x} & \varepsilon_{B_1^1} & 0 & \varepsilon_{t32} & \varepsilon_{t4} & \varepsilon_{t_1^y} & \varepsilon_{t_5^y} \\ \varepsilon_{t_5^x} & \varepsilon_{t_1^x} & 0 & \varepsilon_{B_1^2} & \varepsilon_{t4} & \varepsilon_{t23} & \varepsilon_{t_5^y} & \varepsilon_{t_1^y} \\ \varepsilon_{t_1^y} & \varepsilon_{t_5^y} & \varepsilon_{t32} & \varepsilon_{t4} & \varepsilon_{B_2^1} & 0 & \varepsilon_{t_1^x} & \varepsilon_{t_5^x} \\ \varepsilon_{t_5^y} & \varepsilon_{t_1^y} & \varepsilon_{t4} & \varepsilon_{t23} & 0 & \varepsilon_{B_2^2} & \varepsilon_{t_5^x} & \varepsilon_{t_1^x} \\ \varepsilon_{t23} & \varepsilon_{t4} & \varepsilon_{t_1^y} & \varepsilon_{t_5^y} & \varepsilon_{t_1^x} & \varepsilon_{t_5^x} & \varepsilon_{A_2^1} & 0 \\ \varepsilon_{t4} & \varepsilon_{t32} & \varepsilon_{t_5^y} & \varepsilon_{t_1^y} & \varepsilon_{t_5^x} & \varepsilon_{t_1^x} & 0 & \varepsilon_{A_2^2} \end{pmatrix}, \tag{32}$$

where

$$\varepsilon_{t_1^x} = -2t_1 \cos k_x = -2t_1 \cos\left(\frac{k_{x'}}{\sqrt{2}}\right), \tag{33}$$

$$\varepsilon_{t_1^y} = -2t_1 \cos k_y = -2t_1 \cos\left(\frac{k_{y'}}{\sqrt{2}}\right), \tag{34}$$

$$\varepsilon_{t_5^x} = -2t_5 \cos k_x = -2t_5 \cos\left(\frac{k_{x'}}{\sqrt{2}}\right), \tag{35}$$

$$\varepsilon_{t_5^y} = -2t_5 \cos k_y = -2t_5 \cos\left(\frac{k_{y'}}{\sqrt{2}}\right), \tag{36}$$

$$\varepsilon_{t_{23}} = -2t_2 \cos\left(k_x + k_y\right) - 2t_3 \cos\left(k_x - k_y\right) = -2t_2 \cos\left(\frac{k_{x'}}{\sqrt{2}} + \frac{k_{y'}}{\sqrt{2}}\right) - 2t_3 \cos\left(\frac{k_{x'}}{\sqrt{2}} - \frac{k_{y'}}{\sqrt{2}}\right), \tag{37}$$

$$\varepsilon_{t_{32}} = -2t_3 \cos\left(k_x + k_y\right) - 2t_2 \cos\left(k_x - k_y\right) = -2t_3 \cos\left(\frac{k_{x'} + k_{y'}}{\sqrt{2}}\right) - 2t_2 \cos\left(\frac{k_{x'} - k_{y'}}{\sqrt{2}}\right), \tag{38}$$

$$\varepsilon_{t_4} = -2t_4 \cos\left(k_x + k_y\right) - 2t_4 \cos\left(k_x - k_y\right) = -4t_4 \cos\frac{k_{x'}}{\sqrt{2}} \cos\frac{k_{y'}}{\sqrt{2}}, \tag{39}$$

$$\varepsilon_{t_6} = -2t_6 \cos 2k_x - 2t_6 \cos 2k_y = -2t_6 \cos\sqrt{2}k_{x'} - 2t_6 \cos\sqrt{2}k_{y'} \tag{40}$$

$$\varepsilon_{A_1^1} = \varepsilon_{t_6} + U\left\langle n_{A_1^1 k\downarrow}\right\rangle + U'\left\langle n_{A_1^2 k\downarrow}\right\rangle + (U' - J_H)\left\langle n_{A_1^2 k\uparrow}\right\rangle - \mu, \tag{41}$$

$$\varepsilon_{A_1^2} = \varepsilon_{t_6} + U\left\langle n_{A_1^2 k\downarrow}\right\rangle + U'\left\langle n_{A_1^1 k\downarrow}\right\rangle + (U' - J_H)\left\langle n_{A_1^1 k\uparrow}\right\rangle - \mu, \tag{42}$$

$$\varepsilon_{B_1^1} = \varepsilon_{t_6} + U\left\langle n_{B_1^1 k\downarrow}\right\rangle + U'\left\langle n_{B_1^2 k\downarrow}\right\rangle + (U' - J_H)\left\langle n_{B_1^2 k\uparrow}\right\rangle - \mu, \tag{43}$$

$$\varepsilon_{B_1^2} = \varepsilon_{t_6} + U\left\langle n_{B_1^2 k\downarrow}\right\rangle + U'\left\langle n_{B_1^1 k\downarrow}\right\rangle + (U' - J_H)\left\langle n_{B_1^1 k\uparrow}\right\rangle - \mu. \tag{44}$$

According to the itinerant picture, the interactions between two sets of pockets give rise to a SDW order at the wave vector connecting them with $Q_x = (\pi, 0)$ or $Q_y = (0, \pi)$. For example, as choosing the ordering vector Q_x, the spin configuration has antiparallel spins *along the* X′-direction and parallel spins along the Y′-direction [35]. The antiferromagnetism causes the gapless structure along the X′-direction, and the ferromagnetism opens a gap along the Y′-direction, as shown in **Figure 4(a)**. There are four pockets centered at the Γ point, and two pockets along the Γ − Y′ direction are inequivalent to other two pockets along the Γ − X′ direction. The C_2 symmetry of the Fermi surface resulting from the SDW gap is shown in **Figure 4(b)**.

In the nematic state, the antiparallel spins are along both the X″ *and* Y″ directions. The nematic unit cell is oriented at a 45° with respect to the nonmagnetic unit cell. The nematic Brillouin zone is also a square oriented at a 45° angle with respect to the crystal Brillouin zone. Since the antiferromagnetism is along both the X″ *and* Y″ *directions, there is no gap in the band structure. In addition, as* $M_1 < M_2$, the bands along the Y″ *direction are lifted higher and* cause the bands to be asymmetric with respect to the Γ point. The asymmetric bands result in deformed Fermi surfaces near the Γ point. As shown in **Figure 5(a)**, the blue curve in the band structure forms an elliptic hole-pocket Fermi surface. Furthermore, the elliptic Fermi surface results from the unequal contribution of two orbitals d_{xz} and d_{yz}. The mechanism behind the fluctuations of d_{xz} and d_{yz} orbitals can be understood from an extended RPA approach where d_{xz}, d_{yz}, and d_{xy} orbitals equally contribute to the SDW instability, and in particular the d_{xy} orbital plays a strong role in the nematic instability [30]. The nematic instability breaks the degeneracy of two orbitals d_{xz} and d_{yz} and causes the unequal charge distributions $n_{xz}(k)$ and $n_{yz}(k)$. The C_2-symmetry of the Fermi surface results from the broken degeneracy of two orbitals d_{xz} and d_{yz} (blue and red curves in **Figure 5(b)**).

Recently, Qureshi et al. [53], Wang et al. [54], Steffens et al. [55], and Luo et al. [56] pointed out that in-plane spin excitations exhibit a large gap and indicating that the spin anisotropy is caused by the contribution of itinerant electrons and the topology of Fermi surface. These experiments indicate that the elliptic spin fluctuations at low energy in iron pnictides are mostly caused by the anisotropic damping

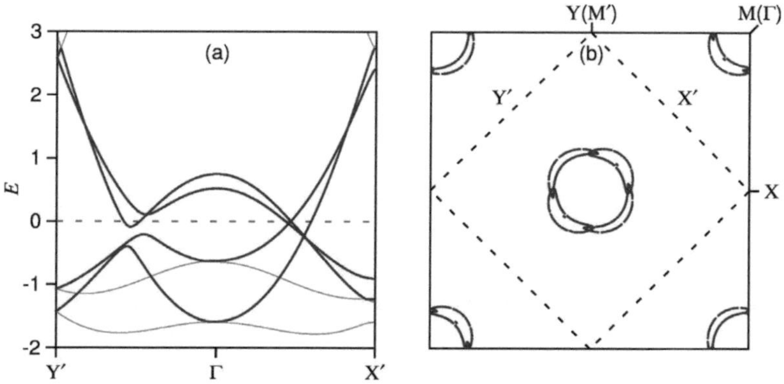

Figure 4.
(color online) (a) and (b) are, respectively, the band structure and the Fermi surface in the stripe SDW state. The Fermi energy (red dashed line) corresponds to the electron filling n = 2.04. X′ and Y′indicate the magnetic Brillouin zone of the stripe SDW state.

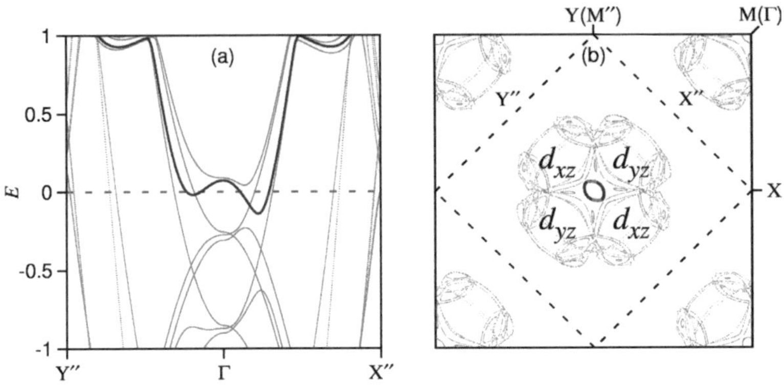

Figure 5.
(color online) (a) and (b) are, respectively, the band structure and the Fermi surface in the nematic state. The asymmetric band (blue color) is responsible for the elliptic Fermi surface around the Γ point. The blue and red solid curves represent the major contributions from the orbitals d_{xz} and d_{yz}, respectively. The chemical potential is chosen as the electron filling $n = 2.1$. X'', and Y'' indicate the magnetic Brillouin zone of the nematic state.

of spin waves within FeAs plane and the topology of Fermi surface. The degeneracy of orbitals will introduce the single-ion anisotropy in spin fluctuations.

3. Visualize nematicity in a lattice

To visualize the nematicity in a lattice, we self-consistently solve the Bogoliubov-de Gennes (BdG) equations for the nematic state in a two-dimensional square lattice:

$$\sum_{jv}\begin{pmatrix} h_{ijuv\uparrow} & \Delta_{ijuv} \\ \Delta^*_{ijuv} & h^*_{ijuv\downarrow} \end{pmatrix}\begin{pmatrix} u^n_{jv\uparrow} \\ v^n_{jv\downarrow} \end{pmatrix} = E_{n\uparrow}\begin{pmatrix} u^n_{iu\uparrow} \\ v^n_{iu\downarrow} \end{pmatrix}, \tag{45}$$

where

$$\mathrm{h}_{\mathrm{iju}\alpha} = \mathrm{t}_{\mathrm{iju}\alpha} + \{-\mu + \mathrm{U}\langle \mathrm{n}_{\mathrm{iu}\beta}\rangle + \mathrm{U}'\langle \mathrm{n}_{\mathrm{iv}\alpha}\rangle + (\mathrm{U}' - \mathrm{J}_{\mathrm{H}})\langle \mathrm{n}_{\mathrm{iv}\beta}\rangle\}\delta_{\mathrm{ij}}, \tag{46}$$

and $U' = U - 2J_H$. The self-consistency conditions are

$$\langle n_{iu\uparrow}\rangle = \sum_n \left|u^n_{iu\uparrow}\right|^2 f(E_n), \tag{47}$$

$$\langle n_{iu\downarrow}\rangle = \sum_n \left|v^n_{iu\downarrow}\right|^2 f(1 - E_n), \tag{48}$$

$$\Delta_{ijuu} = \frac{V}{2}\sum_n u^n_{iu\uparrow} v^{n*}_{iu\downarrow} \tanh\left(\frac{\beta E_n}{2}\right). \tag{49}$$

Here, $f(E_n)$ is the Fermi distribution function.

In **Figure 6**, we show the magnetic configuration in the coexisting state of the nematic order and SC. To view the detail of the structure, the slided profile along the peaks along the x- or y-direction is made (as shown on the sides of M_i in **Figure 6**). There are two sinusoidally modulated magnetizations on each panel. The warm color and cold color modulations represent the spin-up modulation and the spin-down modulation, respectively. The amplitude of each modulation on the left and right panels corresponds to the value of M_1 and M_2. On the left panel, we have $M_1 \cos\left(q_y \cdot r_i\right) = 0.04 \cos\left(\frac{2\pi r_i}{28a}\right)$, and on the right panel, we have $M_2 \cos\left(q_x \cdot r_i\right) = 0.06 \sin\left(\frac{2\pi r_i}{28a}\right)$. Both modulations have the same period $28a$. Since $M_1 < M_2$, the configuration is orthorhombic and breaks the 90° rotational symmetry.

Figure 7 shows the Fourier transformation of the spatial configuration of the nematic fluctuations. Two peaks appearing at $(\pm\pi, 0)$ correspond to the ordering vector Q_x and two peaks exhibiting at $(0, \pm\pi)$ correspond to the ordering vector Q_y. We found that the intensities of peaks associated with the ordering vector Q_x are greater than peaks associated with the ordering vector Q_y, which is due to the unequal value of M_1 and M_2. The intensities of two peaks along the $k_x\,(k_y)$-direction have the same magnitude indicating $\langle\Delta_x\rangle = \langle\Delta_y\rangle = 0$. Moreover, the

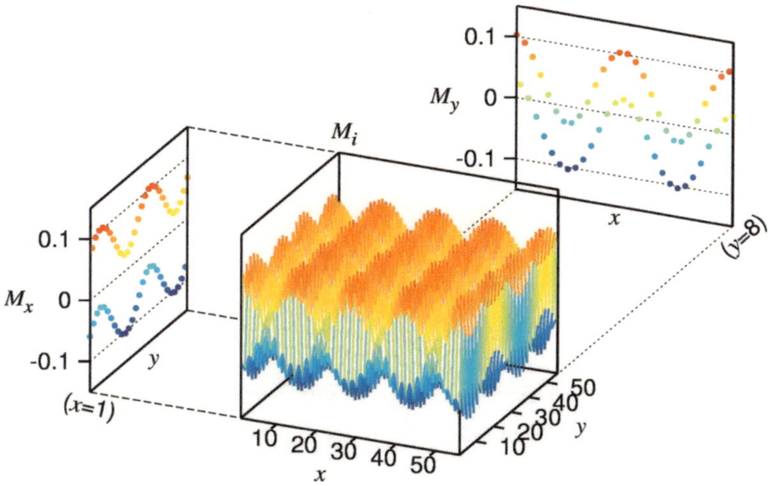

Figure 6.
(color online) the real space configurations of the magnetization M_i are plotted on a 56 × 56 square lattice. The left and the right panels are the sliced profile along the peaks along the y- and x-directions, respectively. Two curves are shown in both panels. The upper and lower curves represent the spin-up and spin-down configurations, respectively.

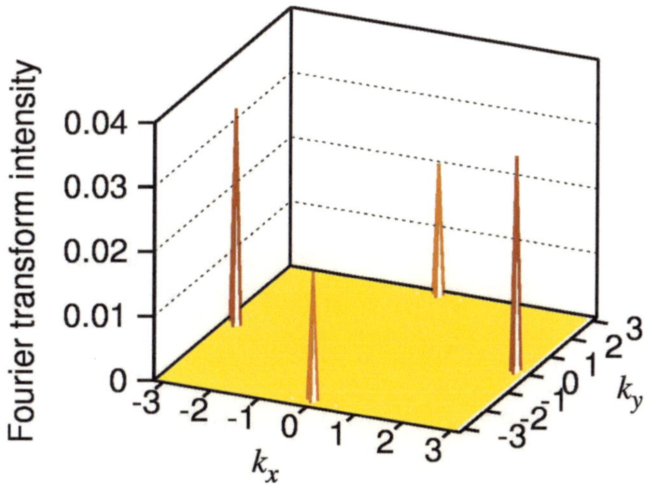

Figure 7.
(color online) the Fourier transformation of the 56 × 56 spatial magnetic configuration.

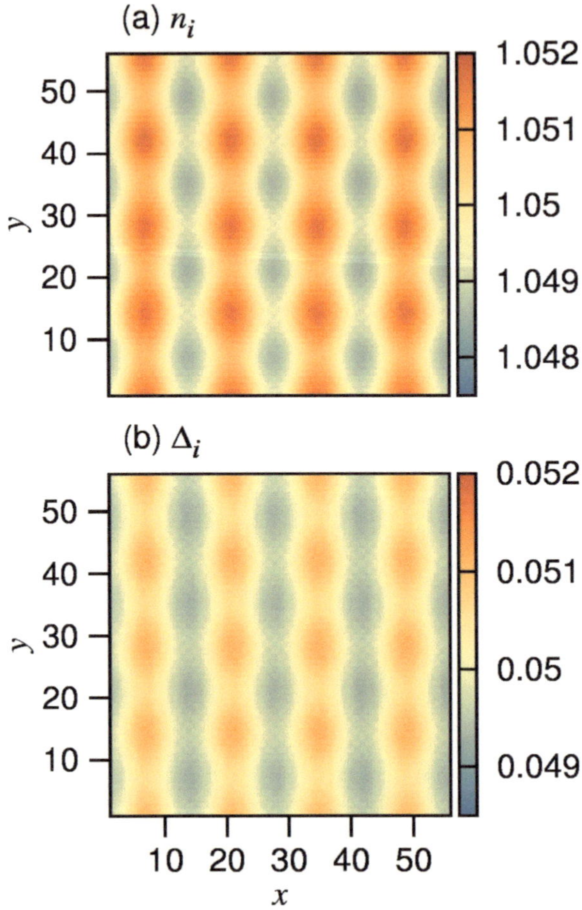

Figure 8.
(color online) (a) the spatial configuration of the electronic charge density n_i. (b) the spatial configuration of the s^{+-}-wave superconducting order parameter Δ_i.

nonequivalence of the intensities between the k_x- and k_y-directions indicates $\langle \Delta_y^2 \rangle > \langle \Delta_x^2 \rangle$. Therefore, the modulated antiparallel spin configuration is the nematic state. These features are preserved even as the SC order is equal to zero. This result is in agreement with the neutron scattering experiments [3, 10].

We further illustrate the electronic charge density $n_i = (n_{i\uparrow} + n_{i\downarrow})$ and the s^{+-}-wave SC order parameter Δ_i as shown in **Figure 8(a)**, **(b)**. Particularly, the nematicity of the spin order induces a modulated charge density wave (CDW) which does not occur in the stripe SDW state. The CDW consists of crisscrossed horizontal and vertical stripes. The amplitudes of the vertical stripes are larger than the horizontal stripes. Therefore, the CDW forms a checked pattern, instead of a checkerboard pattern. The stripes on both x- and y- directions have the same period $14a$ which is the half period of the magnetization.

Moreover, although the checked pattern of the CDW is twofold symmetry, the CDW exhibits a $d_{x^2-y^2}$-symmetry, instead of a d_{xy}-symmetry (diagonal stripes crisscrossed pattern), form factor density wave. The space configuration of the SC order parameter Δ_i shows the same features as the CDW order, as shown in **Figure 8(b)**.

4. The local density of states

The local density of states (LDOS) proportional to the differential tunneling conductance as measured by STM is expressed as

$$\rho_i(E) = -\frac{1}{N_x N_y}\sum_{nu}\left[\left|u^n_{iu\uparrow}\right|^2 f'(E_n - E) + \left|v^n_{iu\downarrow}\right|^2 f'(E_n + E)\right], \tag{50}$$

where $N_x \times N_y = 24 \times 24$ is the size of supercells.

In the striped SDW state, spins are parallel in the y-direction and antiparallel in the x-direction and cause the gap and gapless features in the band structure, respectively. The SDW gap shifts toward negative energy, and the coherence peak at the negative energy is pushed outside the SDW gap and enhanced. The coherence peak at the positive energy is moved inside the SDW gap and suppressed. This is a prominent feature caused by the magnetic SDW order that the intensities of superconducting coherence peaks are obvious asymmetry (as shown in **Figure 9(a)**) [57].

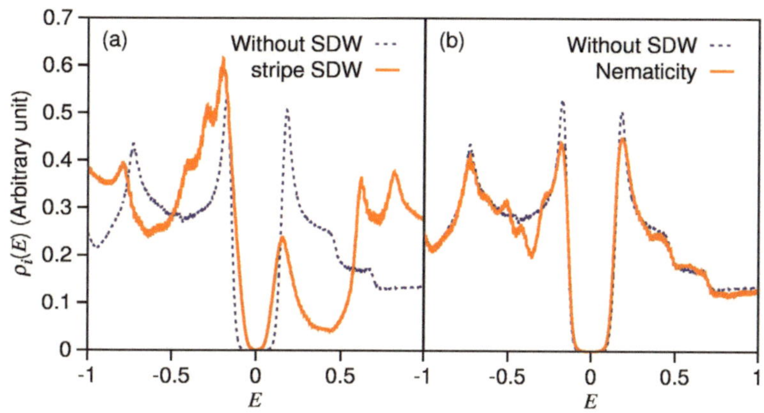

Figure 9.
(color online) the LDOS in the (a) stripe SDW state and (b) nematic state. The dashed (blue) line represents the LDOS without magnetization ($M_i = 0$).

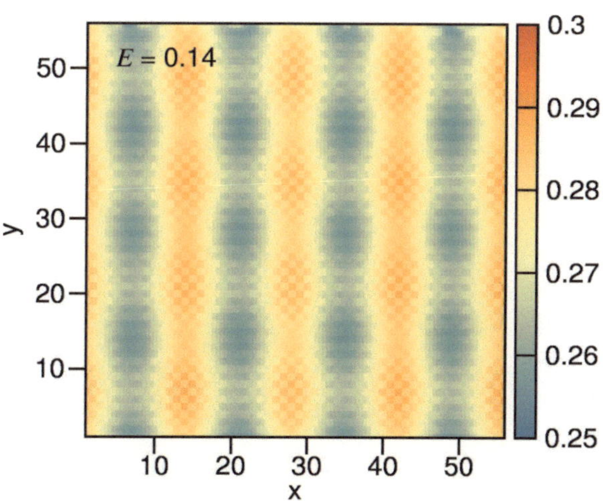

Figure 10.
(Color online) The LDOS map at $E = 0.14$ with $M_i \neq 0$.

In the nematic state, spins are antiparallel in the x- and y-directions leading to a gapless feature in the band structure. The superconducting gap is the only gap that appears in the LDOS. Moreover, comparing to the state without SDW, the competition between the nematic order and the superconducting order causes the slightly suppression of the coherence peaks. The feature of the suppression results in a dip at the negative energy outside the coherence peaks (as shown in **Figure 9(b)**).

Furthermore, **Figure 10** displays a spatial distribution of LDOS, also known as LDOS map, at $E = 0.14$. The LDOS map shows the same features as the charge density distribution at the energy within the coherence peaks. The LDOS map exhibits a checked pattern, twofold symmetric configuration, and $d_{x^2-y^2}$-symmetry form factor density wave. These features have not yet been reported by STM experiments.

It is worth to note that STM measurements by Chuang et al. [5] and Allan [58] reported that the dimension of the electronic nanostructure is around $8a$ and the nanostructure aligns in a unidirectional fashion. The highly twofold symmetric structure of the QPI patterns is represented by using the Fourier transformation of the STS imaging. Moreover, in cuprate, the more advanced measurement of the atomic-scale electronic structure has shown a d-wavelike symmetry form factor density wave [59, 60]. There are four peaks that appear around the center of the momentum space in the QPI patterns. Such an atomic-scale feature in cuprates has not yet been reported in iron pnictides and in the nematic state.

5. Phase diagram

To further verify the spin configuration of the nematic order, a phase diagram is presented in **Figure 11**. In the phase diagram, the stripe SDW order, nematic order, and s^{+-}-wave superconducting order as a function of doping are obtained from the self-consistent calculation to solve the BdG equations.

In the hole-doped region, the magnetization exhibits the stripe SDW order and drops dramatically around $n = 1.85$ and vanishes at $n = 1.80$. In the meantime, the s^{+-}-wave superconducting order reaches its maximal value at $n = 1.85$ and then gradually decreases.

In the electron-doped region, the stripe SDW order (green curve) has its maximal value at $n = 2.00$, then rapidly diminishes in a small region of doping, and finally reaches zero at $n = 2.09$. The superconductivity (blue curve) swiftly increases in a small region from $n = 2.02$ to $n = 2.04$, then reaches its maximal

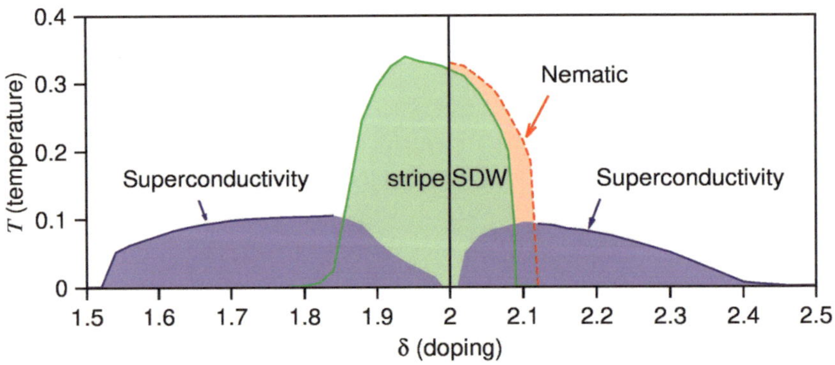

Figure 11.
(color online) the phase diagram of the stripe SDW order (blue), nematic order (green), and superconducting order (red) as a function of doping.

value at $n = 2.10$, and finally gradually decreases to almost zero around $n = 2.40$. The nematic state (red region) is in a small region next to the stripe SDW where the nematic transition line (red dashed curve) tracks closely the stripe SDW transition line. The nematic order is favored to appear in the electron-doped regime, but not the hole-doped regime.

There are two regions where the stripe SDW coexist with the SC and the nematic order coexists with the SC. In the region where the stripe SDW coexist with the SC, the magnetic structure is an orthorhombic uniaxial stripe state. The ordering vector is either Q_x or Q_y implying $\langle \Delta_x \rangle = 0$ or $\langle \Delta_y \rangle = 0$. In the region where the nematic order coexists with the SC, the magnetic structure is a crisscrossed stripe state with twofold symmetry. The ordering vectors are Q_x and Q_y associated with two modulating vectors q_x and q_y implying $\langle \Delta_x \rangle = \langle \Delta_y \rangle = 0$ and $\left\langle \Delta_x^2 - \Delta_y^2 \right\rangle \neq 0$ [61].

It is worth to note that the phase diagram of the electron-doped region is consistent with Figure 1.3 of Kuo's thesis on $Ba(Fe_{1-x}Co_xAs)_2$ [62]. Both figures show the same behavior of the nematic phase. Near the optimally doped region under the superconducting dome, it is mentioned that the long-range nematic order coexists with the superconductivity. However, such results still have not been reported from experiments. In addition, the magnetoresistivity of $Ba_{0.5}K_{0.5}Fe_2As_2$ reported a nematic superconducting state recently and suggested that the hole-doped superconductor is the mixture of *s*-wave and *d*-wave superconducting orders [63]. These results provide a different path to further researches to understand the mechanism of the nematic state in the superconductivity.

6. Conclusions

The two-orbital Hamiltonian used in the iron-based superconductors has always been questioned for its validity. Many studies have approved that a lot of phenomena are attributed to d_{xz} and d_{yz} orbitals. In particular, d_{xz} and d_{yz} orbitals are responsible for the SDW instability.

The stripe SDW order opens a gap in the band structure and deforms the Fermi surface. However, the band structure of the nematic order is gapless, and the Fermi surface is deformed to an ellipse. The mechanism can be understood from the instability of SDW. The nematic order has visualized as a checked pattern formed by a crisscrossed modulated horizontal and vertical stripes. The inequivalent strengths of the horizontal and vertical stripes break the degeneracy of two orbitals d_{xz} and d_{yz} and cause an elliptic Fermi surface. The Fourier transformation of the orthorhombic structure of the magnetization shows two uneven pairs of peaks at $(\pm\pi,0)$ and $(0,\pm\pi)$. Moreover, the LDOS map shows a $d_{x2\text{-}y2}$-symmetry form factor density wave.

Finally, the nematic order is favored to exist in the electron-doped regime, but not the hole-doped regime.

Acknowledgements

HYC was supported by MOST of Taiwan under Grant MOST 107-2112-M-003-002 and National Center for Theoretical Science of Taiwan.

Author details

Hong-Yi Chen
Department of Physics, National Taiwan Normal University, Taipei, Taiwan

*Address all correspondence to: hongyi@ntnu.edu.tw

References

[1] Christianson AD, Goremychkin EA, Osborn R, Rosenkranz S, Lumsden MD, Malliakas CD, et al. Unconventional superconductivity in Ba0.6K0.4Fe2As2 from inelastic neutron scattering. Nature. 2008;**456**:930

[2] Séamus Davis JC, Lee D-H. Concepts relating magnetic interactions, intertwined electronic orders, and strongly correlated superconductivity. PNAS. 2013;**110**:17623

[3] Zhao J, Adroja DT, Yao D-X, Bewley R, Li S, Wang XF, et al. Spin waves and magnetic exchange interactions in CaFe2As2. Nature Physics. 2009;**5**:555

[4] Yi M, Donghui L, Chu J-H, Analytis JG, Sorini AP, Kemper AF, et al. Symmetry-breaking orbital anisotropy observed for detwinned Ba(Fe1-xCox)2As2 above the spin density wave transition. PNAS. 2011;**108**:6878

[5] Chuang T-M, Allan MP, Lee J, Xie Y, Ni N, Bud'ko SL, et al. Nematic electronic structure in the "Parent" State of the iron-based superconductor Ca(Fe1–xCox)2As2. Science. 2010;**327**:181

[6] Chu J-H, Analytis JG, De Greve K, McMahon PL, Islam Z, Yamamoto Y, et al. In-plane resistivity anisotropy in an underdoped iron arsenide superconductor. Science. 2010;**329**:824

[7] Chu J-H, Analytis JG, Press D, De Greve K, Ladd TD, Yamamoto Y, et al. In-plane electronic anisotropy in underdoped Ba(Fe1−xCox)2As2 revealed by partial detwinning in a magnetic field. Physical Review B. 2010;**81**:214502

[8] Shimojima T, Ishizaka K, Ishida Y, Katayama N, Ohgushi K, Kiss T, et al. Orbital-dependent modifications of electronic structure across the Magnetostructural transition in BaFe2As2. Physical Review Letters. 2010;**104**:057002

[9] Tanatar MA, Blomberg EC, Kreyssig A, Kim MG, Ni N, Thaler A, et al. Uniaxial-strain mechanical detwinning of CaFe2As2 and BaFe2As2 crystals: Optical and transport study. Physical Review B. 2010;**81**:184508

[10] Harriger LW, Luo HQ, Liu MS, Frost C, Hu JP, Norman MR, et al. Nematic spin fluid in the tetragonal phase of BaFe2As2. Physical Review B. 2011;**84**:054544

[11] Nakajima M, Liang T, Ishida S, Tomioka Y, Kihou K, Lee CH, et al. Unprecedented anisotropic metallic state in undoped iron arsenide BaFe2As2 revealed by optical spectroscopy. PNAS. 2011;**108**:12238

[12] Kuo H-H, Chu J-H, Riggs SC, Yu L, McMahon PL, De Greve K, et al. Possible origin of the nonmonotonic doping dependence of the in-plane resistivity anisotropy of Ba(Fe1−xTx)2As2(T=Co, Ni and Cu). Physical Review B. 2011;**84**:054540

[13] Kasahara S, Shi HJ, Hashimoto K, Tonegawa S, Mizukami Y, Shibauchi T, et al. Electronic nematicity above the structural and superconducting transition in BaFe2(As1−xPx)2. Nature. 2012;**486**:382

[14] Chu J-H, Kuo H-H, Analytis JG, Fisher IR. Divergent nematic susceptibility in an iron arsenide superconductor. Science. 2012;**337**:710

[15] Fernandes RM, Böhmer AE, Meingast C, Schmalian J. Scaling between magnetic and lattice fluctuations in iron pnictide superconductors. Physical Review Letters. 2013;**111**:137001

[16] Blomberg EC, Tanatar MA, Fernandes RF, Mazin II, Shen B, Wen H-H, et al. Sign-reversal of the in-plane resistivity anisotropy in hole-doped iron pnictides. Nature Communications. 2013;**4**:1914

[17] Mirri C, Dusza A, Bastelberger S, Chu J-H, Kuo H-H, Fisher IR, et al. Nematic-driven anisotropic electronic properties of underdoped detwinned Ba(Fe1−xCox)2As2 revealed by optical spectroscopy. Physical Review B. 2014; **90**:155125

[18] Fernandes RM, Chubukov AV, Schmalian J. What drives nematic order in iron-based superconductors? Nature Physics. 2014;**10**:97

[19] Lv W, Phillips P. Orbitally and magnetically induced anisotropy in iron-based superconductors. Physical Review B. 2011;**84**:174512

[20] Lee W-C, Phillips PW. Non-Fermi liquid due to orbital fluctuations in iron pnictide superconductors. Physical Review B. 2012;**86**:245113

[21] Arham HZ, Hunt CR, Park WK, Gillett J, Das SD, Sebastian SE, et al. Detection of orbital fluctuations above the structural transition temperature in the iron pnictides and chalcogenides. Physical Review B. 2012;**85**:214515

[22] Stanev V, Littlewood PB. Nematicity driven by hybridization in iron-based superconductors. Physical Review B. 2013;**87**:161122(R)

[23] Yamase H, Zeyher R. Superconductivity from orbital nematic fluctuations. Physical Review B. 2013; **88**:180502(R)

[24] Fernandes RM, Chubukov AV, Knolle J, Eremin I, Schmalian J. Preemptive nematic order, pseudogap, and orbital order in the iron pnictides. Physical Review B. 2012;**85**:024534

[25] Hu JP, Xu C. Nematic orders in Iron-based superconductors. Physica C. 2012; **481**:215

[26] Fernandes RM, Schmalian J. Manifestations of nematic degrees of freedom in the magnetic, elastic, and superconducting properties of the iron pnictides. Superconductor Science and Technology. 2012;**25**:084005

[27] Avci S, Chmaissem O, Allred JM, Rosenkranz S, Eremin I, Chubukov AV, et al. Magnetically driven suppression of nematic order in an iron-based superconductor. Nature Communications. 2014;**5**:3845

[28] Xingye L, Park JT, Zhang R, Luo H, Nevidomskyy AH, Si Q, et al. Nematic spin correlations in the tetragonal state of uniaxial-strained BaFe2−xNixAs2. Science. 2014;**345**:657

[29] Zhang W, Park JT, Lu X, Wei Y, Ma X, Hao L, et al. Effect of Nematic order on the low-energy spin fluctuations in Detwinned BaFe1.935Ni0.065As2. Physical Review Letters. 2016;**117**: 227003

[30] Christensen MH, Kang J, Andersen BM, Fernandes RM. Spin-driven nematic instability of the multiorbital Hubbard model: Application to iron-based superconductors. Physical Review B. 2016;**93**:085136

[31] Khalyavin DD, Lovesey SW, Maneul P, Kruger F, Rosenkranz S, Allred JM, et al. Symmetry of reentrant tetragonal phase in Ba1−xNaxFe2As2: Magnetic versus orbital ordering mechanism. Physical Review B. 2014;**90**:174511

[32] Böhmer AE, Hardy F, Wang L, Wolf T, Schweiss P, Meingast C. Superconductivity-induced re-entrance of the orthorhombic distortion in Ba1−xKxFe2As2. Nature Communications. 2015;**6**:7911

[33] Wang X, Kang J, Fernandes RM. Magnetic order without tetragonal-symmetry-breaking in iron arsenides: Microscopic mechanism and spin-wave spectrum. Physical Review B. 2015;**91**: 024401

[34] Gastiasoro MN, Andersen BM. Competing magnetic double-Q phases and superconductivity-induced reentrance of C2 magnetic stripe order in iron pnictides. Physical Review B. 2015;**92**:140506(R)

[35] Pan L, Li J, Tai Y-Y, Graf MJ, Zhu J-X, Ting CS. Evolution of the Fermi surface topology in doped 122 iron pnictides. Physical Review B. 2013;**88**: 214510

[36] Zhao YY, Li B, Li W, Chen H-Y, Bassler KE, Ting CS. Effects of single- and multi-substituted Zn ions in doped 122-type iron-based superconductors. Physical Review B. 2016;**93**:144510

[37] Gao Y, Huang H-X, Chen C, Ting CS, Su W-P. Model of vortex states in hole-doped Iron-Pnictide superconductors. Physical Review Letters. 2011;**106**:027004

[38] Kuroki K, Onari S, Arita R, Usui H, Tanaka Y, Kontani H, et al. Unconventional pairing originating from the disconnected Fermi surfaces of superconducting LaFeAsO1−xFx. Physical Review Letters. 2008;**101**: 087004

[39] Lee PA, Wen X-G. Spin-triplet p-wave pairing in a three-orbital model for iron pnictide superconductors. Physical Review B. 2008;**78**:144517

[40] Daghofer M, Nicholson A, Moreo A, Dagotto E. Three orbital model for the iron-based superconductors. Physical Review B. 2010;**81**:014511

[41] Vorontsov AB, Vavilov MG, Chubukov AV. Interplay between magnetism and superconductivity in the iron pnictides. Physical Review B. 2009; **79**:060508(R)

[42] Maiti S, Fernandes RM, Chubukov AV. Gap nodes induced by coexistence with antiferromagnetism in iron-based superconductors. Physical Review B. 2012;**85**:144527

[43] Parker D, Vavilov MG, Chubukov AV, Mazin II. Coexistence of superconductivity and a spin-density wave in pnictide superconductors: Gap symmetry and nodal lines. Physical Review B. 2009;**80**:100508

[44] Beaird R, Vekhter I, Zhu J-X. Impurity states in multiband s-wave superconductors: Analysis of iron pnictides. Physical Review B. 2012;**86**: 140507(R)

[45] Maisuradze A, Yaouanc A. Magnetic form factor, field map, and field distribution for a BCS type-II superconductor near its Bc2(T) phase boundary. Physical Review B. 2013;**87**: 134508

[46] Gastiasoro MN, Hirschfeld PJ, Andersen BM. Impurity states and cooperative magnetic order in Fe-based superconductors. Physical Review B. 2013;**88**:220509(R)

[47] Zhang D. Nonmagnetic impurity resonances as a signature of sign-reversal pairing in FeAs-based superconductors. Physical Review Letters. 2009;**103**:186402

[48] Tai YY, Zhu J-X, Matthias JG, Ting CS. Calculated phase diagram of doped BaFe2As2superconductor in a C4-symmetry breaking model. Europhysics Letters. 2013;**103**:67001

[49] Castellani C, Natoli CR, Ranninger J. Magnetic structure of V2O3 in the insulating phase. Physical Review B. 1978;**18**:4945

[50] Ghaemi P, Wang F, Vishwanath A. Andreev bound states as a phase-sensitive probe of the pairing symmetry of the iron pnictide superconductors. Physical Review Letters. 2009;**102**: 157002

[51] Zhang X, Oh YS, Liu Y, Yan L, Kim KH, Greene RL, et al. Observation of the Josephson effect in Pb/Ba1−xKxFe2As2 single crystal junctions. Physical Review Letters. 2009;**102**:147002

[52] Seo K, Bernevig BA, Hu J. Pairing symmetry in a two-orbital exchange coupling model of Oxypnictides. Physical Review Letters. 2008;**101**: 206404

[53] Qureshi N, Steffens P, Wurmehl S, Aswartham S, Büchner B, Braden M. Local magnetic anisotropy in BaFe2As2: A polarized inelastic neutron scattering study. Physical Review B. 2012;**86**: 060410(R)

[54] Wang C, Zhang R, Wang F, Luo H, Regnault LP, Dai P, et al. Longitudinal spin excitations and magnetic anisotropy in Antiferromagnetically ordered BaFe2As2. Physical Review X. 2013;**3**:041036

[55] Steffens P, Lee CH, Qureshi N, Kihou K, Iyo A, Eisaki H, et al. Splitting of resonance excitations in nearly optimally doped Ba(Fe0.94Co0.06) 2As2: An inelastic neutron scattering study with polarization analysis. Physical Review Letters. 2013;**110**: 137001

[56] Luo H, Wang M, Zhang C, Lu X, Regnault L-P, Zhang R, et al. Spin excitation anisotropy as a probe of orbital ordering in the paramagnetic tetragonal phase of superconducting BaFe1.904Ni0.096As2. Physical Review Letters. 2013;**111**:107006

[57] Pan L, Li J, Tai Y-Y, Graf MJ, Zhu J-X, Ting CS. Evolution of quasiparticle states with and without a Zn impurity in doped 122 iron pnictides. Physical Review B. 2014;**90**:134501

[58] Allan MP, Chuang T-M, Massee F, Xie Y, Ni N, Bud'ko SL, et al. Anisotropic impurity states, quasiparticle scattering and nematic transport in underdoped Ca(Fe1−xCox) 2As2. Nature Physics. 2013;**9**:220

[59] Fujita K, Hamidian MH, Edkins SD, Kim CK, Kohsaka Y, Azuma M, et al. Direct phase-sensitive identification of a d-form factor density wave in underdoped cuprates. PNAS. 2014;**111**: E3026

[60] Hamidian MH, Edkins SD, Kim CK, Davis JC, Mackenzie AP, Eisaki H, et al. Atomic-scale electronic structure of the cuprate d-symmetry form factor density wave state. Nature Physics. 2016;**12**:150

[61] Chou C-P, Chen H-Y, Ting CS. The nematicity induced d -symmetry charge density wave in electron-doped iron-pnictide superconductors. Physica C. 2018;**546**:61

[62] Kuo H-H. Electronic Nematicity in Iron-Based Superconductors [Thesis]. Stanford University; 2014

[63] Li J, Pereira PJ, Yuan J, Lv Y-Y, Jiang M-P, Lu D, et al. Nematic superconducting state in iron pnictide superconductors. Nature Communications. 2017;**8**:1880